HYPOCRISY++

On Philosophy of
Probability and Sociology
of Ideologies

HYPOCRISY++

On Philosophy of Probability and Sociology of Ideologies

Krzysztof Burdzy

University of Washington, USA

World Scientific

NEW JERSEY • LONDON • SINGAPORE • BEIJING • SHANGHAI • HONG KONG • TAIPEI • CHENNAI

Published by

World Scientific Publishing Co. Pte. Ltd.
5 Toh Tuck Link, Singapore 596224
USA office: 27 Warren Street, Suite 401-402, Hackensack, NJ 07601
UK office: 57 Shelton Street, Covent Garden, London WC2H 9HE

British Library Cataloguing-in-Publication Data
A catalogue record for this book is available from the British Library.

HYPOCRISY++
On Philosophy of Probability and Sociology of Ideologies

ISBN 978-981-98-0547-1 (hardcover)
ISBN 978-981-98-0548-8 (ebook for institutions)
ISBN 978-981-98-0549-5 (ebook for individuals)

For any available supplementary material, please visit
https://www.worldscientific.com/worldscibooks/10.1142/14105#t=suppl

To the memory of Nitka, my cat

Preface

I have published two books on the philosophy of probability. The first one, [1], was devoted to a critique of the two most popular, among scientists, philosophical theories of probability: subjective and frequency. That book also contained my own scientific theory of probability. My second book, [2], presented a general epistemological framework supporting my theory of probability. The present book contains a meta-analysis. A part of it is devoted to the analysis of the intellectual milieu, which allows for two unrealistic philosophical theories to thrive.

My attempts to understand the strange landscape of the philosophy of probability led me to three ideas explaining the situation: hypocrisy++ (a higher level of hypocrisy), manipulation of examples, and the existence of IICs (Insular Ideological Communities). As a bonus, I realized that these phenomena are not limited to the philosophy of probability—they apply to diverse ideologies. None of the three ideas is totally new or revolutionary, but I have not seen a meta-analysis of the philosophy of probability done from this angle.

The book is focused on Bayesian/subjective ideology developed by de Finetti because the frequency philosophy of probability developed by von Mises is practically unknown among scientists. The identification of probability and frequency works well in many cases and is the basis of much of empirical scientific research. It is a naive (in the positive sense) version of the frequency philosophy, and in this form, it is pretty harmless.

The truly significant philosophical claim of von Mises that individual events do not have probabilities seems to be completely unknown and, one can guess, incomprehensible to most scientists.

I reused a chunk of the material from my two previous books to make the present book a stand-alone volume. Much of the old material has been substantially modified. Chapters 4 and 13–17 are new. Chapters 1 and 6 contain significantly revised material from [2]. The remaining chapters, adapted from [2], have been substantially shortened and slightly modified.

I am grateful to the people who have offered their comments and advice over the many years: Itai Benjamini, Erik Björnemo, Nicolas Bouleau, Arthur Fine, Artur Grabowski, Peter Hoff, Wilfrid Kendall, Dan Osborn, Jim Pitman, Jeffrey Rosenthal, Jaime San Martin, Pedro Terán, John Walsh, Natalie Wellen and anonymous referees.

I acknowledge with gratitude the generous support from the Simons Foundation.

Seattle, 2024

Contents

Chapter 1

Introduction

[...] when philosophy makes itself intelligible,
it commits suicide [...] Martin Heidegger[1]

I wish Heidegger had been joking. Maybe he was. No matter his intentions, his words perfectly fit the philosophy of probability. I will argue that its two most popular directions, the subjective theory developed by de Finetti and the frequency theory developed by von Mises, are viable mainly because of confusion, hypocrisy and ignorance permeating and surrounding these philosophical theories.

In the rest of the introduction, I will outline the book's contents, starting from the material adapted from [1] and [2]. The main new contribution is the meta-analysis of the philosophy of probability. I will present some ideas explaining the unbelievable success of some nonsensical theories.

1.1 Philosophy of probability

Two and two make four. Imagine a mathematical theory that says it makes no sense to talk about the result of adding two and two. Imagine another mathematical theory that says that the result of adding two and two is whatever you think it is. Would you consider any of these theories a reasonable foundation of

[1] *Contributions to Philosophy (On Appropriation).* Translated by Richard Polt in *Heidegger. An Introduction*, p. 141.

science? Would you think that they are relevant to ordinary life?

If you toss a coin, the probability of heads is 1/2. According to the frequency philosophy of probability, the probability of heads on a single coin toss makes no sense. According to the subjective philosophy of probability, the probability of heads is whatever you think. Would you consider these theories a reasonable foundation of science? Would you think that they are relevant to ordinary life?

The frequency philosophy of probability is usually considered to be the basis of the "frequency" statistics and the subjective philosophy of probability is often regarded as the basis of the "Bayesian" statistics. According to the frequency philosophy of probability, the concept of probability is limited to long runs of identical experiments or observations and the probability of an event is the relative frequency of the event in the long sequence. The subjective philosophy claims that there is no objective probability, and so probabilities are subjective views; they are rational and beneficial only if they are "consistent," that is if they satisfy the usual mathematical probability formulas.

Von Mises, who created the frequency philosophy, claimed that ([3], p. 11);

> We can say nothing about the probability of death of an individual [within a year] even if we know his condition of life and health in detail.

De Finetti, who proposed the subjective philosophy, asserted that ([4], p. x);

> Probability does not exist.

The standard education in probability and statistics is a process of indoctrination in which students are taught, explicitly or implicitly, that individual events have probabilities, and some methods of computing probabilities are scientific and rational. An alien visiting our planet from a different galaxy would have never guessed from our textbooks on probability and statistics that the two main branches of statistics are related to the philosophical claims cited above.

In a nutshell, both of the two most popular philosophies of probability, frequency and subjective, fail in their own distinct ways. First, both theories are very weak. The frequency philosophy of von Mises, developed in the first half of the twentieth century, provides analysis of long sequences of independent and identical events only. The subjective philosophy of de Finetti (developed in parallel to that of von Mises, more or less) offers an argument supporting the mathematical rules of probability, with no hint on how the rules can be matched with the real world. Second, each of the two philosophical theories failed in a "technical" sense. The frequency theory is based on "collectives," a notion completely abandoned by the scientific community long ago. The subjective theory is based on the "Dutch book" argument, which is no more than a philosophical toy example. Even one of the two failures would be sufficient to disqualify these theories. The double failure makes each of the theories an embarrassment to the scientific community.

My own scientific theory of probability consists of the following laws.

(L1) Probabilities are numbers between 0 and 1 (including 0 and 1), assigned to events whose outcome may be unknown.

(L2) If events A and B cannot happen at the same time, then the probability that A or B will occur is the sum of the probabilities of the individual events, that is, $P(A \text{ or } B) = P(A) + P(B)$.

(L3) If events A and B are physically independent then they are independent in the mathematical sense, that is, $P(A \text{ and } B) = P(A)P(B)$.

(L4) If events A and B are symmetric, then the two events have equal probabilities, that is, $P(A) = P(B)$.

(L5) When an event A is observed, then the probability of B changes from $P(B)$ to $P(A \text{ and } B)/P(A)$.

(L6) An event has probability 0 if and only if it cannot occur. An event has probability 1 if and only if it must occur.

The laws (L1)-(L6) provide a codification of the science of probability at the same level as laws known in some fields of physics, such as thermodynamics or electromagnetism. People familiar with the probability theory at the college level will notice that (L1)-(L6) are a concise summary of the first few chapters of any standard undergraduate probability textbook.

I will show that frequency statistics has nothing (essential) in common with the frequency philosophy of probability and Bayesian statistics has nothing (essential) in common with the subjective philosophy of probability. The two branches of statistics are implicitly based on the laws (L1)-(L6).

1.2 Resonance

The acquisition of information and creating knowledge (this includes both facts and theories) can be divided into two steps. I will call the first step "resonance." This process is very fast in most cases, subconscious and very reliable in a great variety of situations. I guess resonance is not based on logic in any reasonable sense of the word "logic." Resonance is a crude but reasonably reliable filter of information arriving at our senses.

Resonance is fallible in many situations recognized as significant to individual lives, society and science. Logic, probability, induction and all other named and unnamed ingredients of science provide the second filter, much more refined and reliable than resonance. Traditionally, philosophy was focused on the second filter because we have almost no access to resonance via our consciousness. This situation created various misconceptions concerning the sources of reliable truth. One of these is a tendency to ignore resonance even though resonance is at least as important to science and general knowledge as logic. Another common misconception is that resonance (under the name of "subjectivity") is unreliable or not needed. Some other myths go in the opposite direction and invest intuition, subjective opinions and mystical experiences in powers that these sources of opinion and information do not have.

Resonance is a necessary first filter because it is impossible to process all information available to us logically in a timely manner. This should not be interpreted as a claim that humans created resonance deliberately. Quite the opposite, resonance is the result of the blind evolution process selecting the fittest individuals.

A generalized problem of induction is to find a justification for making predictions of future events based on our knowledge of past events. I will argue that the problem of induction is ill-posed in the sense that our "knowledge" of facts is based on the same ontological and epistemological assumptions as our predictions of "unknown" events. The reliability of induction is a law of nature or rather the confluence of several laws of nature, including evolution. These laws of nature are specific to our universe so no general logical justification of induction can exist.

1.3 Sociology of ideologies

Aristotle devised the idea of Five Ws and How: "Who, what, when, where, why, and how?" (see [5]). For the investigation of the failed philosophical theories of probability, the relevant questions seem to be as such:

(i) Why didn't anyone notice the problem?
(ii) Why did de Finetti and von Mises create absurd philosophical theories?
(iii) How is it possible that absurd philosophical theories function as the support of a perfectly good scientific theory—statistics?

The answer to (i) is that some people did notice the problem. See Sec. 6.4.4 for some critical opinions about the subjective philosophy of probability. The book [6], among other publications, contains a critique of the frequency philosophy.

Concerning (ii), de Finetti and von Mises created very weak theories to avoid the discussion of the objective probability of a single event (see Sec. 2.6).

My answer to (iii) is the principal new contribution of this book. I identify three phenomena as the culprits: hypocrisy++, missing examples, and Insular Ideological Communities (IIC). I will show that these phenomena exist in several branches of intellectual activity besides the philosophy of probability.

I will define, explain and illustrate hypocrisy++, a higher level of hypocrisy, in Chap. 13.

The title of Chap. 14 refers to "Schrödinger's cat," a thought experiment representing an apparent paradox in quantum physics. Philosophy of probability is characterized by the lack of explicit, detailed and clear analysis of real and imaginary experiments, real-life random phenomena, and human opinions and decisions. One of the goals of this volume is to present and analyze taboos in the mainstream philosophy of probability.

Chap. 15 will analyze IICs, intellectual environments where irrational theories can thrive.

The last (but not least) of the main themes of the book will be a review of two recent publications: [6] and [7]. The authors of these books declared themselves to be supporters of the subjective philosophy of probability. I will show that the books are permeated by objectivist ideas.

PART 1

Philosophy and Probability

Chapter 2

Main Philosophies of Probability

My general classification of the main philosophies of probability is borrowed from [8–10]. Some authors pointed out that even the classification of probability theories is rife with controversy, so the reader should not be surprised to find a considerably different list in [11].

I will present only these versions of popular philosophical theories of probability, which I consider clear. In other words, this chapter plays a double role. It is a short introduction to the philosophy of probability for those unfamiliar with it. It is also my attempt to clarify the essential claims of various theories. I think I am faithful to the spirit of all theories I present, but I will make little effort to present the nuances of their various interpretations. In particular, I will discuss only the version of the frequency theory that claims that probability is not an attribute of a single event. This is because I believe the need for the concept of a "collective" used in this theory is justified only when we adopt this assumption. Similarly, I do not see how one can construct a coherent and convincing theory that includes objective and subjective probabilities.

I will pay much more attention to the subjective and frequency theories than other theories because these two theories are widely believed to be the foundation of modern statistics and other applications of probability. I will discuss less popular philosophies of probability first.

2.1 The classical theory

Traditionally, the birth of the modern mathematics-based probability theory is dated back to the correspondence between Pierre de Fermat and Blaise Pascal in 1654. They discussed a problem concerning dice posed by Chevalier de Mere, a gambler. Some calculations of probabilities can be found in earlier books (see Chap. 1 of [9] for more details).

The "classical" definition of probability gives a mathematical recipe for calculating probabilities in highly symmetric situations, such as tossing a coin, rolling a die or playing cards. It does not seem to be concerned with the question of the "nature" of probability. In 1814, Laplace stated the definition in these words (English version of [9], p. 17):

> The theory of chance consists in reducing all the events of the same kind to a certain number of cases equally possible, that is to say, to such as we may be equally undecided about in regard to their existence, and in determining the number of cases favorable to the event whose probability is sought. The ratio of this number to that of all the cases possible is the measure of this probability, which is thus simply a fraction whose numerator is the number of favorable cases and whose denominator is the number of all the cases possible.

Since the definition applies only to those situations in which all outcomes are (known to be) equally "possible," it does not make it clear what one should think about an experiment with only a single toss, or an experiment with a deformed coin—does the concept of probability apply to these situations at all? The classical definition seems to be circular because it refers to "equally possible" cases—this presumably means "equally probable" cases—and so probability is defined using the notion of probability.

The "classical philosophy of probability" is a modern label. That "philosophy" was a practical recipe, not a conscious attempt to create a philosophy of probability, unlike all other philosophies reviewed below, which were developed in the twentieth century, partly in parallel.

2.2 The logical theory

The "logical" theory of probability maintains that probabilities are numbers representing relations between propositions. They are weak forms of logical implication, intuitively speaking. According to this theory, the study of probability is equivalent to the study of a formal language. John Maynard Keynes and, later, Rudolf Carnap were the most prominent representatives of this philosophical view. Their main books were [12] and [13]. The version of the theory advocated by Keynes allowed for non-numerical probabilities. The logical theory is based on the "principle of indifference", which asserts that equal probabilities should be assigned to alternatives for which no reason is known to be different. The principle is also known as the "principle of insufficient reason" and goes back to Laplace. It was criticized as early as 1854 by Boole, who thought that our ignorance is not a good justification for assigning equal probabilities to events.

The principle of indifference does not have a unique interpretation. If you toss a deformed coin twice, what is the probability that the results will be different? There are four possible results: HH, TH, HT and TT (H stands for heads, T stands for tails). The principle of indifference suggests that all four results are equally likely, so the probability that the results will differ is $1/2$. A generalization of this claim to a large number n of tosses says that all sequences of outcomes are equally likely. A simple mathematical argument then shows that the tosses are (mathematically) independent and the probability of heads is $1/2$ for each toss. Since this conclusion is not palatable, Keynes and Carnap argued that the probability that the results of the first two tosses will be different should be taken as $1/3$. This claim and its generalizations are mathematically equivalent to choosing the "uniform prior" in the Bayesian setting. In other words, we should assume that the tosses are independent and identically distributed with the probability of heads that is itself a random variable—a number chosen uniformly from the interval $[0, 1]$.

The logical theory seems almost unknown to mathematicians, probabilists, and statisticians. The emphasis on the logical aspect of probability seems to miss the real difficulties with this concept. Statisticians and scientists seem to be quite happy with Kolmogorov's mathematical theory as the formal basis of probability. Almost all of the controversy is concerned with the implementation of the formal theory in practice.

The boundaries between different philosophies are not sharp. For example, Carnap believed in two different concepts of probability, one appropriate for logic, and another one appropriate for physical sciences.

The logical theory is also known as a "necessary" or "a priori" interpretation of probability.

2.3 The propensity theory

The term "propensity theory" is applied to recent philosophical theories of probability, which consider probability to be an objective property of things or experiments just like mass or electrical charge. Karl Popper developed the first version of the propensity theory (see [14]).

The following example illustrates a problem with this interpretation of probability. Suppose a company manufactures identical computers in plants in Japan and Mexico. The propensity theory does not provide a convincing interpretation of the statement "This computer was made in Japan with probability 70%," because it is hard to imagine what physical property this sentence might refer to.

Popper advanced another philosophical idea, namely, that one can falsify probabilistic statements that involve probabilities very close to 0 or 1. He said ([14], Sec. 68, p. 202);

> The rule that extreme improbabilities have to be neglected [...] agrees with the demand for *scientific objectivity*.

This idea is implicit in all theories of probability, and one form of it was stated by Cournot in the first half of the nineteenth century (quoted from [15], p. 585):

> If the probability of an event is sufficiently small, one should act in a way as if this event will not occur at a solitary realization.

A version of Cournot's principle and Popper's idea appears in this book as (L6)—an essential part of my own theory.

2.4 The subjective theory

Two people arrived independently at the idea of the subjective theory of probability in 1930s, Frank Ramsey and Bruno de Finetti. Ramsey did not live long enough to develop his theory fully, so de Finetti was the founder and the best-known representative of this school of thought.

The "subjective" theory of probability identifies probabilities with subjective opinions about unknown events. This idea is deceptively simple. First, the word "subjective" is ambiguous. Second, one must explain why the mathematical probability theory should be used if there is no objective probability.

De Finetti emphatically denied the existence of any objective probabilistic statements or objective quantities representing probability. He summarized this in his famous saying, "Probability does not exist."

In de Finetti's theory, "subjective" means "non-existent," while the popular (but false) interpretation of "subjective" is "informally measured."

The radical idea of de Finetti was that probability statements could not be verified at all—this is the fundamental meaning of subjectivity in his philosophical theory. This idea can be presented as a great triumph of thought or as a great failure. If you admire de Finetti, you may emphasize the simplicity and elegance of his solution to the verification problem. If you are his detractor, you may say that de Finetti could not solve a philosophical problem, so he tried to conceal his failure by declaring that the problem was ill-posed.

In de Finetti's theory, probability statements are "subjective" in the sense that "no probability statement is verifiable or falsifiable in any objective sense." Actually, according to de

Finetti, probability statements are not verifiable in any sense, "subjective" or "objective." In his theory, when new information is available, it is not used to verify any probability statements made in the past. The subjective probabilities do not change at all—the only thing that happens is that one starts to use different probabilities, based on the old *and* new information. This does not affect the original probability assignments, except that they become irrelevant for making decisions—they are not falsified, according to de Finetti. Observing the occurrence of an event or its complement cannot falsify or verify any statement about its probability.

De Finetti's philosophical theory is devoid of any means whatsoever of verifying any probability statement. This extreme position, not universally adopted by subjectivists, is an indispensable element of the theory; I will discuss this further in Chap. 6. A good illustration of this point is the following commentary of de Finetti on the fact that beliefs in some probability statements are common to all scientists, and so they seem to be objective and verifiable (quoted from [9], p. 70):

> Our point of view remains in all cases the same: to show that there are rather profound psychological reasons which make the exact or approximate agreement that is observed between the opinions of different individuals very natural, but there are no reasons, rational, positive, or metaphysical, that can give this fact any meaning beyond that of a simple agreement of subjective opinions.

The subjective theory is rich in ideas—no sarcasm is intended here. I will discuss some of these ideas: the "Dutch book" argument, the axiomatic system for the subjective theory, the identification of probabilities and decisions, and the Bayes theorem.

Probability does not exist in an objective sense, according to the subjective theory, so why should we use the probability calculus at all? One can justify the application of the mathematical theory of probability to subjective probabilities using a Dutch book argument. A Dutch book will be formed against me if I place various bets in such a way that no matter which events occur and which do not occur, I will lose some money.

One can prove rigorously that it is possible to make a Dutch book against a person if and only if the "probabilities" used by the person are not "consistent", that is, they do not satisfy the usual formulas of the mathematical probability theory.

I will illustrate the idea of a Dutch book with a simple example. Consider an experiment with only three possible mutually exclusive outcomes A, B and C. For example, these events may represent winners of a race with three runners. The mathematical theory of probability requires the probabilities of A, B and C to be non-negative and to add up to 1; that is, $P(A) + P(B) + P(C) = 1$. The complement of an event A is traditionally denoted A^c; that is, A^c is the event that A did not occur, and the mathematical theory of probability requires that $P(A^c) = 1 - P(A)$. Suppose that I harbor "inconsistent" views, for example, my personal choice of probabilities is $P(A) = P(B) = P(C) = 0.9$, so that $P(A) + P(B) + P(C) > 1$. Since I am 90% sure that A will happen, I am willing to pay someone \$0.85, assuming that I will receive \$1.00 if A occurs (and nothing otherwise). Placing this bet seems to be rational because the expected gain is equal to $\$0.15 \cdot P(A) - \$0.85 \cdot P(A^c) = \$0.15 \cdot 0.9 - \$0.85 \cdot 0.1 = \$0.05$, a strictly positive number. A similar calculation shows that I should also accept two analogous bets, with A replaced by B and C. If I place all three bets, I will have to pay $\$0.85 + \$0.85 + \$0.85 = \$2.55$. Only one of the events A, B or C may occur. No matter which event occurs, A, B or C, I will receive the payoff equal to \$1.00 only. In each case, I am going to lose \$1.55. A Dutch book was formed against me because I did not follow the usual rules of probability, that is, I used "probabilities" that did not satisfy the condition $P(A) + P(B) + P(C) = 1$.

Consistency protects me against creating a situation resulting in certain loss so I have to use the mathematics of probability in my judgments, the subjective theory advises. Note that the claim here is not that inconsistency will necessarily result in a Dutch book situation (in a given practical situation, there may be no bets offered to me), but that consistency protects me against a Dutch book situation under all circumstances.

The essence of the Dutch book argument is that one can achieve a deterministic and empirically verifiable goal using probability calculus, without assuming anything about the existence of objective probabilities.

The subjective theory of probability is sometimes introduced using an axiomatic system, as in [16] or [17]. This approach gives the subjective theory of probability the flavor of a mathematical (logical, formal) theory. The postulates are intuitively appealing, even obvious—just what one would expect from axioms.

If a theory of probability is developed in the decision theoretic context, it is clear that one needs to deal with the question of the "real" value of money and the value of non-monetary rewards, such as friendship. An accepted way to deal with the problem is to introduce a utility function. A one dollar gain typically has a different utility for a pauper compared to a millionaire. It is commonly assumed that the utility function is increasing and convex, that is, people prefer to have more money than less money (you can always give away the unwanted surplus), and the subjective satisfaction from the gain of an extra dollar is smaller and smaller as your fortune grows larger and larger.

The ultimate subjectivist approach to probability is to start with a set of axioms for rational decision making in the face of uncertainty and derive the mathematical laws of probability from these axioms. This approach was developed in [18], but it was based in part on the von Neumann-Morgenstern theory (see [17], Chap. 14). If one starts from a number of quite intuitive axioms concerning decision preferences, one can show that there exists a probability measure P and a utility function such that a decision $\mathcal{A}$ is preferable to $\mathcal{B}$ if and only if the expected utility is greater if we take action $\mathcal{A}$ rather than action $\mathcal{B}$, assuming that we calculate the expectation using P. If a probability distribution and a utility function are given, then the decision-making strategy that maximizes the expected utility satisfies the axioms proposed by Savage. Deriving probabilities from decision preferences does not guarantee that probability values are related in any way to reality. One can only prove theoretically a formal equivalence of a consistent decision strategy and a probabilistic view of the world.

The subjective theory is implemented in Bayesian statistics in a very specific way. The essence of statistics is the analysis of data, so the subjective theory has to supply a method for incorporating data into a consistent set of opinions. On the mathematical side, the procedure is called "conditioning," that is, if some new information is available, the holder of a consistent set of probabilistic opinions should start using the probability distribution *conditional* on the extra information. The mathematical theorem that shows how to calculate the conditional probabilities is called the Bayes theorem. A consistent set of opinions held before the data are collected is called the "prior distribution," or simply the "prior," and the probability distribution obtained from the prior and the data using the Bayes theorem is called the "posterior distribution," or the "posterior."

2.5 The frequency theory

The development of the foundations of the mathematical theory of probability at the end of the seventeenth century was related to observations of the stability of relative frequencies of some events in gambling. In the middle of the nineteenth century, John Venn and other philosophers developed a theory identifying probability with frequency. At the beginning of the twentieth century, Richard von Mises formalized this idea using the concept of a collective. A collective is a long (ideally, infinite) sequence of isomorphic events. Examples of collectives include casino-type games of chance, repeated measurements of the same physical quantity such as the speed of light, and measurements of a physical quantity for different individuals in a population, such as blood pressure of patients in a hospital. Von Mises defined a collective using its mathematical properties. For a sequence of observations to be a collective, the relative frequency of an event must converge to a limit as the number of observations grows. The limit is then identified with the probability of the event.

Many observations related to weather show seasonal patterns and the same is true for some business activities. Von Mises did not consider such sequences to be collectives, so he imposed an extra condition that relative frequencies of the event should be equal along subsequences chosen without prophetic powers.

Von Mises ruled out applications of probability outside the realm of collectives (p. 28 of [3]):

> It is possible to speak about probabilities only in reference to a properly defined collective.

Examples of collectives given by von Mises are very similar to those used to explain the notions of i.i.d. (independent identically distributed) random variables, or exchangeable sequences (see Chap. 3). Both the definition of an i.i.d. sequence and the definition of an exchangeable sequence include, among other things, the condition that the probabilities of the first two events in the sequence are equal. In von Mises' theory, individual events do not have probabilities. Hence, collectives could not be defined in the same way as i.i.d. or exchangeable sequences. Instead, he used the principle of "place selection," that is, he required that frequencies were stable along all subsequences of a collective chosen without prophetic powers.

Quite often, the frequency theory of von Mises and the subjective theory of de Finetti are portrayed as the antipodal options in the philosophical discourse. This is sometimes expressed by labeling the two theories as "objective" and "subjective." In fact, the fundamental claims of both ideologies make them sister theories. De Finetti claimed that the probability of an event is not measurable in any objective sense, and so did von Mises. These negative claims have profound consequences in both scientific and philosophical arenas. Von Mises argued that there is an objectively measurable quantity that can be called "probability," but it is an attribute of a long sequence of events, not an event. De Finetti thought that it could be objectively proved that probabilities are beneficial if they are assigned in a "consistent" way. Hence, both philosophers agreed that there are no objective probabilities of events, but there are some objectively justifiable scientific practices involving probability.

2.6 Comparison of philosophical theories of probability

A striking feature of the current intellectual atmosphere is that the two most popular philosophical theories in statistics—frequency and subjective—are the only theories that deny that the concept of probability applies to individual events (in an objective way). The philosophical ideas of von Mises and de Finetti were as revolutionary as those of Einstein. Einstein and other physicists forced us to revise our basic instincts concerning the relationship between the observed and the observer, the role of space and time, the relationship between mass and energy, the limits of scientific knowledge, etc. The idea that events do not have probabilities is equally counterintuitive.

2.6.1 *The meaning of objectivity*

A mainstream meaning of "objective" is that something is true without human involvement. The Sun is objectively bigger than the Moon. The Sun would still be bigger than the Moon if people did not exist. See also Sec. 4.3.

Different philosophies of probability connect mathematical theory and reality in different ways. Every philosophy expects people to believe in the objectivity of certain facts and properties. Philosophical theories assume that humans are able to observe objective facts and reliably communicate them to others. In this sense, all philosophical theories are partly objective. Implicitly, none of the theories claims that people can assess probability directly and objectively, although we unquestionably have instinctive feelings about probability. In some cases, it is asserted that objective probabilities can be derived from objective facts and properties; for example, in logical theory, objective probabilities can be derived from objective observations of symmetry. In the extreme case of the subjective theory, objective probabilities cannot be derived at all.

(i) In the logical theory, probabilities are numbers associated with propositions. In other words, mathematical events are replaced with propositions in this theory. To apply this theory, one must be able to recognize past events and then express the observations as propositions. Crucially, one must be able to recognize the lack of information and symmetry to apply the principle of indifference.
(ii) In the subjective theory, mathematical events are associated with real events. Probabilities are numbers associated with events. One has to be able to recognize mutually exclusive events. One must recognize if an event occurred or not for conditioning, i.e., to apply the Bayes formula.
(iii) In the frequency theory, probabilities are numbers associated with collectives. To apply the theory, one has to be able to recognize collectives, and whether an event occurred (to determine observed frequencies).
(iv) In my own theory (see Chap. 9), probabilities are assigned to events. People are expected to be able to recognize mutually exclusive events, symmetrics, physical independence, and whether an event occurred.

2.7 Peaceful coexistence?

Some philosophers believe that various interpretations of probability can coexist either in the sense that they are applicable in different situations or that they can be mixed; see [19–21] and the references therein. As an example of such an approach, consider the questions that form the starting point of the analysis in [21, p. 4]:

> (i) Given the data, what should we believe, and to what degree?
> (ii) Do the data provide evidence for model $M1$ against an alternative model $M2$, and how much?

(iii) Given the data what should we do?
(iv) Given the data, what is the evidence for the greater predictive accuracy of one model against its alternative?

I consider (i)-(iv) a philosophical program orthogonal to my own. Comparing (i)-(iv) to (a)-(d) in the previous section elucidates the difference. I am trying to describe which facts and properties are (or should be) considered "objective" in the sense of being amenable to reliable assessment and communication to other people. My program has a theoretical side but is also an empirical program—it is my goal to describe probability theory as it is practiced. In my framework, objective and subjective philosophical theories of probability cannot be mixed because either a property such as symmetry or an object such as a collective can be effectively recognized or not, according to a given philosophy. There is no room for compromise. For example, a philosophical theory that recognizes the applications of imperfect symmetries in probability must be prepared to explain their utility. Symmetries cannot be recognized in some practical situations, but this is irrelevant—the philosophical theory that recognizes the uses of symmetries in probability must analyze them no matter whether they are perfect or imperfect, frequent or rare.

[19, p. 89] distinguishes between "Neyman-Pearson formulation" and "evidential formulation" of hypothesis testing. Once again, this train of thought seems to be orthogonal to my program.

2.8 Is there life beyond von Mises and de Finetti?

The foundations of probability look entirely different from different perspectives. I am not even discussing disagreements between philosophers on the nature of probability. Different accounts of the foundations of probability offer strikingly different descriptions of the most significant directions in and contributions to this part of philosophy. Among all philosophical theories of probability, I have focused almost entirely on two theories, those of von Mises and de Finetti. One of the top results of the online search for "philosophy of probability" is an article [22], which presents a rainbow of philosophical ideas. Von Mises and de Finetti are featured in that article, but they by no means dominate the review. So why do I focus on von Mises and de Finetti?

The main reason is that I would like to understand the foundations of statistics. The two main branches of statistics, frequency and Bayesian, seem most strongly connected with the philosophies of von Mises and de Finetti. One of the main theses of this book is that there is practically no relationship between the two branches of statistics and the two philosophical theories. However, I do not know of any other philosophical theories of probability that would match statistics better in the minds of statisticians and other users of statistics and probability.

A good reason to limit the discussion to the theories of von Mises and de Finetti is that they dominate certain areas of the public discussion. Although publications, such as [8, 22, 23], prove that professional philosophers have a broad view of their field, clear signs are showing that much of the scientific and general public does not know much or does not care about probabilistic philosophical ideas other than "frequency" and "subjective." I will support this claim with several examples. The textbook [24] does not focus on just two philosophers, de Finetti and von Mises, but Chap. 13–19 present a sympathetic view of subjectivism and frequentism, leaving no doubt in student's mind that these are the two main directions in the philosophy of probability. The titles of the chapters in [25] contain only three names—those of Kolmogorov, von Mises and de Finetti. A recent book [26] on "philosophy of chance" is very different from the mainstream monographs on the philosophy of probability. Its review of the main philosophical theories of probability is limited to those of von Mises and de Finetti (Sec. 6.7–6.8). Two popular undergraduate textbooks, [27] and [28], discuss only frequency and subjective interpretations of probability.

This handful of examples was enough to convince me that

my main intellectual adversaries in the area of philosophy of probability are de facto von Mises and de Finetti. I did not find any other philosophical theory of probability more attractive than those of von Mises and de Finetti. Many theories sound more reasonable than these two, but they fail to answer crucial questions to my satisfaction.

Chapter 3

Mathematical Methods of Probability and Statistics

I present a review of some mathematical methods of probability and statistics used in the philosophical arguments in this book. This short review is not a substitute for a solid course in probability. Good textbooks at the undergraduate level are [27–29].

3.1 Probability

The mathematics of probability is based on Kolmogorov's axioms. The fully rigorous presentation of the axioms requires some definitions from the "measure theory," a field of mathematics. This material is not needed in this book, so I will present the axioms in an elementary way. No matter how complicated, any probabilistic model is represented by a space of all possible outcomes Ω. The individual outcomes ω in Ω can be very simple (for example, "heads," if you toss a coin) or very complicated (a single outcome ω may represent temperatures at all places around the globe over the next year). Individual outcomes may be combined to form events. If you roll a die, individual outcomes ω are numbers $1, 2, \dots, 6$, that is $\Omega = \{1, 2, 3, 4, 5, 6\}$. The event "even number of dots" is represented by a subset of Ω, specifically, by $\{2, 4, 6\}$. Every event has a probability; that is, probability is a function that assigns a number between 0 and 1 (0 and 1 are not excluded) to every event. If you roll a "fair" die then all outcomes are equally probable, that is, $P(1) = P(2) = \cdots = P(6) = 1/6$. Kolmogorov's axioms put only two restrictions on probabilities. First, $P(\Omega) = 1$. Second, if events $A_1, A_2, \ldots, A_n$ are disjoint, that is, at most one of them can occur, then the probability that at least one of them will occur is the sum of probabilities of $A_1, A_2, \ldots, A_n$. In symbols,

$$P(A_1 \text{ or } A_2 \text{ or } \ldots \text{ or } A_n) = P(A_1) + P(A_2) + \cdots + P(A_n).$$

Kolmogorov's axioms include an analogous statement for a countably infinite sequence of mutually exclusive events—this is called σ-additivity or countable additivity.

A curious feature of Kolmogorov's axiomatic system is that it does not include at all the notion of independence. We call two events (mathematically) independent if the probability of their joint occurrence is the product of their probabilities, in symbols, $P(A \text{ and } B) = P(A)P(B)$. The intuitive meaning of independence is that the occurrence of one of the events does not give any information about the possibility of occurrence of the other event.

If a quantity X depends on the outcome ω of an experiment or observation, then we call it a random variable. For example, if the experiment is a roll of two dice, the sum of dots is a random variable. If a random variable X may take values $x_1, x_2, \ldots, x_n$ with probabilities $p_1, p_2, \ldots, p_n$ then the number $EX = p_1x_1 + p_2x_2 + \cdots + p_nx_n$ is called the expected value or expectation of X. Intuitively speaking, the expectation of X is the (weighted) average, mean or central value of all possible values, although each of these descriptions is questionable. The expected value of the number of dots on a fair die is $(1/6)(1) + (1/6)(2) + \cdots + (1/6)(6) = 3.5$. Note that the "expected value" of the number of dots is not expected at all because the number of dots must be an integer.

The expectation of $(X - EX)^2$, that is, $E((X - EX)^2)$ is called the variance of X and denoted $\text{Var}X$. Its square root is called the standard deviation of X and denoted σ_X, that is $\sigma_X = \sqrt{\text{Var}X}$. It is much easier to explain the intuitive meaning of standard deviation than that of variance. Most random variables take values different from their expectations. The standard deviation represents a typical difference between the value taken by the random variable and its expectation. The strange definition of the standard deviation, via variance and square root, has excellent theoretical support—a mathematical result known as the Central Limit Theorem, to be reviewed next.

3.1.1 *Law of Large Numbers, Central Limit Theorem and Large Deviations Principle*

A sequence of random variables $X_1, X_2, X_3, \ldots$ is called i.i.d. if these random variables are independent and have identical distributions.

The Strong Law of Large Numbers says that if $X_1, X_2, X_3, \ldots$ are i.i.d. and EX_1 exists then the averages $(X_1+X_2+\cdots+X_n)/n$ converge to EX_1 with probability 1 when n goes to infinity.

The Weak Law of Large Numbers asserts that if $X_1, X_2, X_3, \ldots$ are i.i.d. and EX_1 exists then for every $\varepsilon > 0$ and $p < 1$ we can find n so large that

$$P(|(X_1 + X_2 + \cdots + X_n)/n - EX_1| < \varepsilon) > p.$$

A random variable Y is said to have the standard normal distribution if $P(Y < y) = (1/\sqrt{2\pi}) \int_{-\infty}^{y} \exp(-x^2/2)dx$. At the intuitive level, the distribution of possible values of a standard normal random variable is represented by a bell-shaped curve centered at 0.

Suppose that $X_1, X_2, X_3, \ldots$ are i.i.d., the expectation of any of these random variables is μ and its standard deviation is σ. The Central Limit Theorem says that for large n, the normalized sum $(1/(\sigma\sqrt{n}))\sum_{k=1}^{n}(X_k - \mu)$ has a distribution very close to the standard normal distribution.

Roughly speaking, the Large Deviations Principle (LDP) says that under appropriate assumptions, observing a value of a random variable far away from its mean has a probability much smaller than a naive intuition might suggest. For example, if X has the standard normal distribution, the probability that X will take a value greater than x is of order $(1/x)\exp(-x^2/2)$ for large x. The probability that the standard normal random variable will take a value 10 times greater than its standard deviation is about 10^{-38}. The Central Limit Theorem suggests that the Large Deviations Principle applies to sums or averages of sequences of i.i.d. random variables. It does, but the precise formulation of LDP will not be given here. The LDP-type estimates are not always as small as the above example might suggest.

3.1.2 *Exchangeability and de Finetti's theorem*

A permutation π of a set $\{1, 2, \dots, n\}$ is any one-to-one function mapping this set into itself. A sequence of random variables $(X_1, X_2, \dots, X_n)$ is called exchangeable if it has the same distribution as $(X_{\pi(1)}, X_{\pi(2)}, \dots, X_{\pi(n)})$ for every permutation π of $\{1, 2, \dots, n\}$. Informally, $X_1, X_2, \dots, X_n$ are exchangeable if for any sequence of possible values of these random variables, any other ordering of the same values is equally likely.

Recall that a sequence of random variables $(X_1, X_2, \dots, X_n)$ is called i.i.d. if these random variables are independent and have identical distributions. A celebrated theorem of de Finetti says that an infinite exchangeable sequence of random variables is a mixture of i.i.d. sequences. For example, for any given infinite exchangeable sequence of random variables taking values 0 or 1, one can generate a sequence with the same probabilistic properties by first choosing randomly, in an appropriate way, a number p in the interval $[0, 1]$, and then generating an i.i.d. sequence whose elements X_k take values 1 with probability p and 0 with probability $1 - p$.

Deform a coin and encode the result of the k-th toss as X_k, that is, let $X_k = 1$ if the k-th toss is heads and $X_k = 0$ otherwise. Then $X_1, X_2, X_3, \dots$ is an exchangeable sequence. Some probabilists and statisticians consider this sequence to be i.i.d. with "unknown probability of heads."

3.2 Frequency statistics

Statistics is concerned with the analysis of data, although there is no unanimous agreement on whether this means "inference," that is, the search for the truth, or making decisions, or both.

One of the methods of frequency statistics is estimation—I will explain it using an example. Suppose that you have a deformed coin and you would like to know the probability p of heads (this formulation of the problem contains an implicit assumption that the probability p is objective). We can toss the coin n times and encode the results as a sequence of numbers (random variables) $X_1, X_2, \dots, X_n$, with the convention that $X_k = 1$ if the result of the k-th toss is heads and $X_k = 0$ otherwise. Then we can calculate $\overline{p} = (X_1+X_2+\dots+X_n)/n$, which is an "estimator" of p. The estimator $\overline{p}$ is our guess about the true value of p. One of its good properties is that it is "unbiased," that is, its expectation is equal to p. The standard deviation of $\overline{p}$ is $\sqrt{npq}$.

Another procedure used by frequency statisticians is hypothesis testing. Consider the following drug-testing example. Suppose that a new drug is expected to give better results than an old drug. Doctors adopt (temporarily) a hypothesis H (often called a "null hypothesis") that the new drug is *not* better than the old drug and choose a *level of significance*, often 5% or 1%. They then give one of the drugs to a group of patients and the other drug to another group of patients. When the results are collected, the probability of the observed or a "more extreme" result is calculated, assuming that the hypothesis H is true. If the probability is smaller than the significance level then the "null" hypothesis H is rejected and the new drug is declared to be better than the old drug.

On the mathematical side, hypothesis testing proceeds along slightly different lines. Usually, at least two hypotheses are considered. Suppose that you can observe a random variable X whose distribution is either P_0 or P_1. Let H_0 be the hypothesis that the distribution is in fact P_0, and let H_1 be the hypothesis that the distribution of X is P_1. An appropriate number c is found, corresponding to the significance level. When X is observed and its value is x, the ratio of probabilities $P_0(X = x)/P_1(X = x)$ is calculated. If the ratio is less than c, then the hypothesis H_0 is rejected, and otherwise, it is accepted. The constant c can be adjusted to make one of the two possible errors small: rejecting H_0 when it is true or accepting it when it is false.

I will outline the idea of a "confidence interval." Suppose a scientist wants to find the value of a physical quantity θ. Assume that he has a measuring device that does not generate systematic errors; that is, the errors do not tend to be mostly positive or negative. Suppose that the measurements are $X_1, X_2, \dots, X_n$. The average of these numbers, $\overline{X}_n = (X_1+X_2+\dots+X_n)/n$, can be taken as an estimate of θ. The empirical standard deviation $\sigma_n = \sqrt{(1/n)\sum_{k=1}^n (X_k - \overline{X}_n)^2}$ is a measure of accuracy of the estimate. If the number of measurements is large, and some other assumptions are satisfied, the interval $(\overline{X}_n - \sigma_n, \overline{X}_n + \sigma_n)$ covers the true value of θ with probability equal to about 68%. If the length of the interval is increased to 4 standard deviations, that is, if we use $(\overline{X}_n - 2\sigma_n, \overline{X}_n + 2\sigma_n)$, the probability of coverage of the true value of θ becomes 95%.

3.3 Bayesian statistics

Bayesian statistics derives its name from the Bayes theorem. Here is a very simple version of the theorem. Let $P(A \mid B)$ denote the probability of an event A given the information that an event B occurred. Then $P(A \mid B) = P(A \text{ and } B)/P(B)$. Suppose that events A_1 and A_2 cannot occur simultaneously, but one of them must occur. The Bayes theorem is the following formula,

$$P(A_1 \mid B) = \frac{P(B \mid A_1)P(A_1)}{P(B \mid A_1)P(A_1) + P(B \mid A_2)P(A_2)}. \tag{3.1}$$

Intuitively, the Bayes theorem is a form of a retrodiction, that is, it gives the probability of one of several causes (A_1 or A_2), given that an effect (B) has been observed.

One of the simplest examples of the Bayesian method is the analysis of tosses of a deformed coin. A popular Bayesian model assumes that coin tosses are exchangeable. According to de Finetti's theorem, this is mathematically equivalent to the assumption that there exists an unknown number Θ (a random variable), between 0 and 1, representing the probability of heads on a single toss. If we assume that the value of Θ is θ, then the sequence of tosses is i.i.d. with the probability of heads on a given toss equal to θ. The Bayesian analysis starts with a *prior* distribution of Θ. A typical choice is the uniform distribution on $[0, 1]$, that is, the probability that Θ is in a given subinterval of $[0, 1]$ of length r is equal to r. Suppose that the coin was tossed n times and k heads were observed. The Bayes theorem can be used to show that, given these observations and assuming the uniform prior for Θ, the *posterior* probability of heads on the $(n + 1)$-st toss is $(k + 1)/(n + 2)$. Some readers may be puzzled by the presence of constants 1 and 2 in the formula—one could expect the answer to be k/n. If we tossed the coin only once and the result was heads, then the Bayesian posterior probability of heads on the next toss would be $(k + 1)/(n + 2) = 2/3$; this seems to be much more reasonable than $k/n = 1$.

Chapter 4

Philosophy

4.1 What is philosophy?

What is philosophy? Which philosophical theories are valuable? The weight of these questions is crushing. Nevertheless, I have to answer these questions from my perspective because I will argue that two specific philosophical theories are complete intellectual failures.

There was no clear separation between philosophy, science and mathematics in antiquity, although we could impose our current perspective on ancient contributions. Philosophy seems best defined as whatever remained when science and mathematics were removed from ancient philosophy.

From the perspective of this book, it is best to consider mathematics, science and philosophy as communities. Mathematicians are a community that agrees on what constitutes rigorous mathematical proof. The quality of mathematical results is judged in two ways. The first one is conspicuously subjective—some mathematicians consider some theorems elegant or significant. The second quality criterion is less subjective (but not completely objective)—some mathematical results are held in high regard for their applicability in science, engineering, etc.

Science is a community of people who agree on what constitutes empirical proof. Like in mathematics, the quality of scientific results may be judged purely subjectively. However, the relationship between scientific theories and reality is the primary concern. Good theories are expected to have solid empirical support, be accurate, wide-ranging, explanatory, etc.

Philosophy is a community of people who agree on what constitutes philosophy. A traditional point of view is that philosophy deals with fundamental questions and concepts, such as existence, truth and God. But I would de-emphasize the subject matter. Philosophers observe the real universe, think about it, and develop questions, answers, and concepts inspired by the observations. What I expect from significant philosophical writing are ideas that would not occur to an ordinary person or even to a highly educated person occupied with other concerns. I do not ask a philosophical theory to be true or empirically supported. I want a philosophical theory to be interesting. I want to feel that I learned something I could not have invented myself.

A specific vein in philosophy and humanities is to describe, analyze and explain widespread human beliefs. This may seem less challenging than finding the truth about the fundamental components of reality. However, experience shows that describing an ideology such as religion is not any easier than creating a theory of "existence."

I will discuss a few philosophical claims and theorics to illustrate my point of view.

4.1.1 *Motion does not exist*

Zeno of Elea invented several paradoxes that supposedly showed motion does not exist. From the current perspective, this claim was an almost complete intellectual failure.

Let me start with "almost." Zeno linked motion, a physical process, to mathematical concepts of sequence and series. Although the link is weak, the idea of the mathematization of motion can be recognized as a significant step in the right direction from the current perspective.

But things went only down from there. Zeno's paradoxes were based on the impossibility of summing infinite series. Zeno could have concluded that motion does not exist or that infinite series may have finite sums. His intuition failed him. Much later, Galileo, Newton, Einstein (and many others) said much more interesting things about motion. The method of exhaustion developed by Archimedes was a precursor of series summability theory, not fully developed until the 17th century.

Over two millennia passed, and we see that Zeno's claim that motion does not exist did not result in any follow-up theory that would be interesting or significant. The claim was a philosophical dead end.

4.1.2 *Physical universe does not exist*

Solipsism is the claim that the physical universe does not exist. It is impossible to refute this claim. The problem is that it has monumental consequences that nobody can fully embrace, either intellectually or practically. Someone noticed that for a solipsist, it is OK to kill his own mother. In general, a solipsist is not bound by any ethical (or scientific, for that matter) standards. While this observation has some value, it resonates much less than Dostoevsky's famous remark that "If God does not exist, then everything is permitted." Solipsism leads nowhere because it has already jumped to the very end.

4.1.3 *There is no objective truth*

Relativism comes in different forms and degrees—see [30]. The claim that there is no objective truth is its extreme version—parts of [31] come very close to this claim. I offer a short review of this book below to show the inevitable logical consequences of strict relativism and partly for the reader's amusement.

Feyerabend did everything he could to annoy and offend mainstream scientifically minded readers. He mentioned "special professions such as science or prostitution" (p. 217). He praised the Chinese Communist Party for suppressing Western medicine (pp. 50–51). He claimed there is no rational reason why a person should not walk out of a window on the fiftieth floor (p. 221). He called the philosophy of science a "bastard subject"

(p. 301). He proposed to teach "myths of 'primitive' societies" in parallel with science (p. 308). On p. 299, Feyerabend wrote

> The separation between state and church must therefore be complemented by the separation between state and science.
>
> We need not fear that such a separation will lead to a breakdown of technology. [...] We shall develop and progress with the help of the numerous *willing* [emphasis in original] slaves in universities and laboratories who provide us with pills, gas, electricity, atom bombs, frozen dinners [...].

Note that Feyerabend did not see any rational reason why jumping from a fiftieth-floor window would kill you. Still, he somehow knew that willing slaves at universities would provide electricity.

At the other extreme, Feyerabend offered a painstakingly detailed analysis of Galileo's manipulation of telescopic observations. After reading his pontifications on the separation of state and science and jumping from a skyscraper, one has to wonder whether there are any reasons to believe Feyerabend when he said that Galileo existed, that the telescope is an optical instrument and that the Moon in the sky is a real object and not a figment of imagination. Feyerabend's relativism was so extreme that it was more amusing than annoying. Feyerabend was skeptical only when it suited him.

One could adopt a mild form of relativism, saying that all statements mix the subjective and objective. This I agree with. Relativists must choose between a powerful but uninteresting claim that there is no objective truth and a palatable observation that all claims have subjective ingredients. The latter version of the theory is a mundane mainstream philosophy without much depth or luster.

Relativists use their grandiose claims to justify very narrow, politically correct claims or causes fashionable on the left, like pseudoscience in postcolonial countries. The monumental consequences of their general claims are of little interest to their proponents. The same applies to the subjective philosophy of probability. The earthshaking claim that "probability does not exist" is needed for only one extremely narrow purpose—to justify using supposedly subjective priors in Bayesian statistics.

4.1.4 *God does not exist*

This section's main purpose is to highlight the vast difference between investigations of God and those of religion. Atheism goes back to antiquity. Since then, there have been times and locations when and where there was almost universal consensus on God's existence and His nature, but the question of God's existence was never fully resolved. Certainly, it will not be resolved to the satisfaction of all philosophers any time soon.

While it is a viable philosophical position to say that God does not exist, to claim that religion does not exist would be silly. An atheist philosopher has a challenge ready for her—to describe and explain religion. Marx offered this diagnosis: "Religion is the opium of the people." In Marxist philosophy, religion is an ideological weapon used by the ruling class to suppress the working class. It is irrelevant here whether Marxist claims are true or false. What matters is that a social phenomenon needs an explanation, even if the philosopher believes that the central concept of the social phenomenon is an illusion.

Probability is a social phenomenon. It begs for a philosophical explanation, even if one believes objective probability does not exist.

4.2 Applied philosophy

Some philosophical theories can be implemented and verified within our standard intellectual and practical framework. Some cannot.

Zeno's "Motion does not exist" is impossible to interpret or implement in everyday practical life. Any serious attempt in this direction would be classified as mental illness. We do not have the intellectual tools to analyze such experiments sympathetically within the ordinary intellectual framework.

Skepticism: It is impossible to be consistently skeptical about everything. Any serious attempt in this direction would be classified as mental illness. We do not have the intellectual tools to analyze such experiments sympathetically within the ordinary intellectual framework.

Relativism: It is impossible to consider all truths as totally subjective consistently. Any serious attempt in this direction would be classified as mental illness. We do not have the intellectual tools to analyze such experiments sympathetically within the ordinary intellectual framework.

Solipsism can be interpreted as a license to do anything and everything because you affect only your thoughts, and it is your choice how to affect them. Any serious attempt in this direction would be classified as mental illness. However, both the followers of solipsism and observers can collect data on the results of a practical implementation of solipsism. The data can be analyzed. In this sense, experimental solipsism could be scientific.

De Finetti's and von Mises' philosophical theories contain elements (Dutch book and frequency) that can be easily interpreted and implemented. The results can be recorded. The data can be analyzed using standard intellectual tools. Hence, de Finetti's and von Mises' philosophical theories have substantial scientific components.

4.3 Enforced objectivity

No matter what philosophers declare about objective truth and existence, every society enforces some beliefs. They vary widely from society to society—religion is the primary example. But some of the enforced objectivism seems to be universal—here are some of its elements:

(i) People can recognize objects and facts. People know that other people can recognize objects and facts. People know what facts other people can recognize.

(ii) People know the rules of deductive and inductive reasoning. People know that other people can and do apply them.

(iii) People believe they can communicate reliably and effectively with others.

Beliefs (i)-(iii) have practical consequences:

(a) People delegate tasks to others.
(b) People use information about facts collected and theories developed by others.
(c) People establish and enforce rules, regulations and laws governing the society.

The prevalence of enforced objectivity does not prove anything. But it presents two philosophical tasks: to describe enforced objective beliefs and explain their roots.

4.3.1 *Applied subjectivity*

The word "subjective" has multiple meanings, so we cannot declare what is and is not subjective. However, it is important to record some practical aspects of subjectivity.

Subjective opinions about beauty, elegance, tasty food, etc., are used and useful. Still, they cannot be counted upon because they differ from an individual to an individual, from a community to a community, from a historical context to another historical context. More importantly, they can change in time and be manipulated. Even more importantly, from an ethical point of view, it is accepted that subjective opinions can be manipulated, for example, via advertising. The message in an advertisement can purport to be objective (say, "our produce is healthy") or openly subjective (say, "our clothing is elegant"). But no matter what the character of an ad is, its unabashed purpose is to influence perceptions.

Religion is an interesting example of subjective-objective ambivalence. According to [32], apostasy is currently criminalized in about ten countries, all of them Muslim. While this in no way proves the objective truth of Islam, it gives Islam the status accorded to objective truths, for example, scientific truths, in those countries. On the other hand, we find the United States, where the separation of state and church is strictly enforced, giving religion the status of a subjective opinion. At the same time, God's existence is considered an objective truth by a large part of the American population.

I will show that the concept of probability is treated fundamentally differently from the mainstream subjective concepts in Chap. 6.

4.4 How to eat your cake and have it too

A recurring theme in this book is the discrepancy between philosophical claims and their applications. The phenomenon has two versions. The philosophical theory makes weak claims in the first version, but the corresponding practical applications go much further. This is because a weak theory is easy to justify. Applications require whatever it takes to achieve success. Nobody will give up a method that "works" just because philosophers could not develop a solid justification. This type of discrepancy is exemplified by the subjective and frequency philosophical theories of probability. The principle of induction is another well-known example, although philosophers have tried hard to find a justification for this principle.

At the other end of the spectrum are philosophical theories that make grandiose claims but apply them in very limited circumstances. The overgeneralized claims are needed either to make an impression of intellectual courage ("there is no objective truth") or to address an emotional need ("God is omnibenevolent"). The audience seems to be tolerant towards the obvious hypocrisy.

4.5 Dreams

The rise of some philosophical theories, including the subjective and frequency philosophical theories of probability, can be partly attributed to several philosophical and psychological dreams dating back to antiquity. Here are some examples.

(i) The irresistible appeal of utopia: Plato's ideal state, Biblical paradise and heaven, communist state.
(ii) The irresistible appeal of simplicity: Euclid's axioms, Democritus' atoms, Newton's laws of motion, minimalism in art, (some) monastic life, (some) consumer design (iPhone).
(iii) The irresistible appeal of the ultimate answer: Pythagorean numbers, Euclid's axioms, 19th-century physics (Lord Kelvin's claims), modern physics (string theory).
(iv) The irresistible appeal of atoms: Democritus' atoms, Euclid's axioms, 20th-century physics, string theory.
(v) The irresistible appeal of magic: magic, religion, quantum mechanics, relativity.
(vi) The irresistible appeal of mystery: religion, quantum mechanics.
(vii) The irresistible appeal of paradox: One God and Holy Trinity, wave-particle duality in quantum mechanics.

Let us have a closer look at the dream of simplicity. Ancient Greeks believed that everything comprised four elements: earth, water, air, and fire. This theory was suggested around 450 BC and later supported and added to by Aristotle. Euclid based his geometry on axioms. Democritus believed in atoms. Newton invented the laws of motion. Each one of these dreams was quashed. There are more than four elements, and they have isotopes. The currently accepted axiomatic system for mathematics (so-called "ZFC") is non-intuitive; Gödel's theorems show its limitations. Atoms consist of protons, neutrons and electrons; these particles consist of quarks, and quarks might consist of strings. Einstein's physics is much more complicated than Newton's.

The dreams persist. The subjective philosophy of probability tries to derive probability from elementary assumptions. However, the reality is uncooperative—a realistic theory requires strong assumptions.

4.6 Scientists cannot understand philosophy

I am far from suggesting that scientists are not intelligent enough to understand philosophy. I claim that scientists are trained in a mode of thinking that makes it virtually impossible for them to understand philosophy if their standard intellectual habits are not supplanted with an effort to learn a new way of thinking.

Philosophers often make contradictory claims. One reason is that philosophy is not written to the rigorous standards of mathematical logic, so it is harder to detect inconsistencies. Contradictions are much less damaging to philosophical theories than mathematical or scientific ones. Scientists are confused when confronted with illogical philosophical claims. They may miss contradictions and do not know how to react when they find them.

Scientists take facts, deductive and inductive logic for granted. None of this is taken for granted by philosophers.

Scientists use only those terms that are interpersonal. Philosophers often develop their own vocabulary and, more generally, language. A part of being a philosopher is to try to understand other philosophers. Some philosophers claim effective translation is impossible (see [33]). German words are routinely used in English philosophical texts because they cannot be faithfully translated.

Much of this book is devoted to analyzing the scientific and philosophical clash of two intellectual cultures in the foundations of probability.

Chapter 5

Frequency Philosophy of Probability

Von Mises did not contribute anything interesting or significant to the philosophy of probability. The idea of identifying frequency and probability was not new and can be traced back to Bernoulli at the beginning of the eighteenth century. The concept of a "collective" invented by von Mises is awkward, unusable and long abandoned. The claim that individual events do not have probabilities is incomprehensible to most people. Von Mises did not detect patterns in conscious or subconscious methods of assigning probabilities to individual events. Hence, he could not describe and analyze these patterns, even to only dismiss them.

The frequency theory illustrates well a natural tension between philosophy and science. In one area of intellectual activity, the weakest possible claims are the most convenient, while in the other, the strongest possible claims are the most practical. I will argue that there is an unbridgeable gap between the philosophical theory of collectives and the needs of science.

5.1 Von Mises' brand of skepticism

Recall that, according to von Mises (p. 28 of [3]),

> It is possible to speak about probabilities only in reference to a properly defined collective.

This should be interpreted as saying that probability is an attribute of a "collective" and not of an event. A collective is an infinite sequence of observations such that the relative frequency of an event converges to the same number along every subsequence chosen without prophetic powers. The common limit is called the probability (of the event in this collective).

5.1.1 *The smoking gun*

The quote of von Mises presented in Sec. 5.1 is subject to interpretation, just like everything else in philosophy. I will argue that the concept of a collective necessarily leads to the radical interpretation of von Mises' theory. Collectives are equivalent to physical evidence in a court trial. The concept of a collective is a conspicuous element of the frequency theory, and there is no justification for its use except the radical interpretation of the theory.

Examples of real collectives given by von Mises are of the same type as the ones used to illustrate the ubiquitous concept of independent identically distributed (i.i.d.) random variables. On the mathematical side, i.i.d. sequences are much more convenient than collectives. The mathematical concept of i.i.d. random variables was already known in the nineteenth century, even if this name was not always used. Why is it that von Mises chose to formalize a class of models of real phenomena using a considerably less convenient mathematical concept? The reason is that the definition of an i.i.d. sequence of random variables $X_1, X_2, \ldots$ includes, among other things, a statement that the following two events $\{X_1$ is equal to $0\}$ and $\{X_2$ is equal to $0\}$ have equal probabilities (I have chosen 0 as an example; any other value would work as well). According to von Mises' philosophy, a single event, even if it is part of a well-defined collective, does not have its individual probability. In other words, no theoretical or empirical method could determine whether $P(X_1 = 0) = P(X_2 = 0)$. According to von Mises, the notion of an i.i.d. sequence involves non-existent quantities that is, probabilities of individual events.

The fact that collectives are the most prominent part of von

Mises' theory proves that the only interpretation of this theory that is compatible with the original philosophical idea of von Mises is that individual events do not have probabilities.

5.1.2 *Inconsistencies in von Mises' theory*

The claim of von Mises quoted at the beginning of Sec. 5.1 refers to collectives. What constitutes a collective? Are imaginary collectives acceptable from the scientific point of view? The difficulties with von Mises' philosophical theory are well illustrated by inconsistencies in his book [3]. His philosophical ideas require that only "real" collectives are considered. When he applies his philosophical ideas to science, "imaginary" collectives are used.

On p. 9 of [3] we find,

> Our probability theory has nothing to do with questions such as: 'Is there a probability of Germany being at some time in the future involved in a war with Liberia?'

On p. 10, von Mises explained that "unlimited repetition" was a "crucial" characteristic of a collective and gave real life examples of collectives with this feature, such as people buying insurance. He also stated explicitly that

> The implication of Germany in a war with the Republic of Liberia is not a situation which repeats itself.

We could easily imagine a long sequence of planets such as the Earth, with countries such as Germany and Liberia. And we could imagine that the frequency of wars between the pairs of analogous countries on different planets is stable. Von Mises ignored such imaginary collectives. He tacitly considered them to be irrelevant and useless.

Later in the book, von Mises discussed hypothesis testing. On p. 156, he said that hypothesis testing could be approached using the Bayes method. This takes us back to pp. 117–118 of his book. There, he constructed a collective based on the observed data. For example, in the case when the data were observations of a Bernoulli sequence (that is, every outcome was either a "success" or "failure") of length n, and we observed n_1 successes, he constructed a collective using a "partition." That is, he considered a long sequence of data sets, such that in every case, the ratio n_1/n was the same number a. In practice, this construction yields a purely imaginary collective. Except for a handful of trivial applications of statistics, data sets never repeat themselves in real life, even if we look only at "sufficient" statistics (that is, essential general numerical characteristics of the data sets). Hence, von Mises used imaginary collectives in the scientific context.

The inconsistencies in von Mises' book are a case of hypocrisy alluded to in Sec. 4.4.

5.2 Collectives in real life

The concept of a "collective" invented by von Mises is an awkward attempt to formalize the idea of repeated experiments or observations. Two alternative ways to formalize this idea are known as an i.i.d. (independent identically distributed) sequence and "exchangeable" sequence, the latter favored by de Finetti. Exchangeability is a form of symmetry—according to the definition of an exchangeable sequence, any rearrangement of a possible sequence of results is as probable as the original sequence. The idea of an i.i.d. sequence stresses independence of one experiment in a series from another experiment in the same series, given the information about the probabilities of various results for a single experiment. In interesting practical applications, this information is missing, and then, by de Finetti's theorem, an i.i.d. sequence can be equivalently thought of as an exchangeable sequence (see Sec. 3.1.2).

A fundamental problem with collectives is that they would be very hard to use if anybody ever tried to use them. Scientists have to analyze data collected in the past and also to make predictions. For the concept of a collective to apply to the past data, a scientist must be able to recognize a collective effectively. The definition of a collective suggests that one can determine the lack of patterns in the data either at the intuitive level, directly examining the sequence, or using some more formalized but practical procedure. I will discuss scientific methods of detecting patterns below. On the informal side, I do not think people can effectively determine whether a sequence contains non-i.i.d. patterns. Convincing support for my position was provided by a "java applet" prepared by Susan Holmes. The program generated two binary sequences, one simulating i.i.d. events and the other one representing a (non-trivial) Markov chain, that is, dependent events. Many people found it hard to guess whether an unlabeled sequence was i.i.d. or non-i.i.d.

The definition of a collective requires that for a given event, the relative frequency of that event in the sequence (collective) converges, and the same is true for every subsequence of the collective chosen without clairvoyant powers (the limit must always be the same). I believe that people can effectively recognize symmetries and independence via resonance (to be discussed in Sec. 7.2). Our ability to recognize the lack of patterns along "all" subsequences, postulated by von Mises, is an illusion.

The requirement that the relative frequencies have the same limits along "all" subsequences is especially hard to interpret if one has a finite (but possibly long) sequence. In this case, we necessarily have limits 1 and 0 along some subsequences, and it is hard to find a good justification for eliminating these subsequences from our considerations. The reason for the requirement that the limit is the same along all subsequences is to disallow sequences that contain patterns, such as seasonal or daily fluctuations. For example, temperatures at a given location show strong daily and seasonal patterns, so temperature readings do not qualify as a collective. Surprisingly, this seemingly philosophically intractable aspect of the definition of a collective turned out to be tractable in practice in quite a reasonable way. Some of the most important tools used in modern statistics and other sciences are random number generators. These are either clever algebraic algorithms (generating "pseudo-random" numbers) or electronic devices generating random numbers (from thermal noise, for example). From the practical point of view, it is crucial to check that a given random number generator does not produce a sequence that contains patterns, so there is a field of science devoted to the analysis of random number generators. The results seem to be very satisfactory because most statisticians and scientists can find random number generators sufficiently devoid of patterns to meet their needs. In this special sense, von Mises is vindicated—it is possible to check in practice if a sequence is a collective. However, widely used and accepted methods of checking whether a sequence is "truly" random, such as George Marsaglia's battery of tests [34], do not even remotely resemble von Mises' idea of checking the frequency of an event along every subsequence.

Mathematics is used in science to reduce the number of required measurements and to make predictions, among other things. A scientist makes a few measurements and then uses mathematical formulas appropriate for a given science to find values of some other quantities. If we adopt the frequency view of probability, the only predictions offered by this theory are predictions involving limits of long-run relative frequencies. According to the frequency theory, even very complex mathematical results in probability theory should be interpreted as statements about long-run frequencies for large collections of events within the same collective. In most applications of probability, this is unrealistic. The frequency view of the probability theory as a calculus for certain classes of infinite sequences is purely abstract and has very few real applications.

5.3 Collectives and symmetry

A scientific theory has to be applicable in the sense that its laws have to be formulated using terms that correspond to real objects and quantities observable in some reasonable sense. There is more than one way to translate the theory of collectives into an implementable theory. If we use a collective as an observable, we will impose a heavy burden on all scientists because they will have to check for the lack of patterns in all potential collectives. This is done for random number generators out of necessity and in some other practical situations when the provenance of a sequence is not fully understood. But to impose this requirement on all potential collectives would halt the science. An alternative way is to identify a collective with an exchangeable sequence. The invariance under permutations (the defining feature of an exchangeable sequence) can be ascertained directly in many practical situations—this eliminates the need for testing for patterns. This approach is based on symmetry, and so it implicitly refers to (L4) and, more generally, to (L1)-(L6). Hence, it is impossible to implement the concept of a collective, or the concept is redundant.

There is another, closely related, reason why the concept of a collective is almost useless without (L1)-(L6). Typically, when a scientist determines a probability by performing a large number of experiments or collecting a large number of observations, she wants to apply this knowledge in some other context—one could even say that this is the essence of science. Consider the following routine application of statistics. Suppose a group of 1,000 patients was given a drug and improvement occurred in 65% of cases. A statistician can make a prediction that if 2 million people afflicted by the same ailment take this drug, about 1.3 million will experience improvement in their condition. The statistician must be able to select a part of the general population to which the prediction can be applied. The statistician cannot observe limits along any subsequences until the drug is widely used. Making a prediction requires that the statistician uses symmetry to identify the relevant part of the population—here, applying symmetry means identifying people with similar medical records. One can analyze the performance of the drug *a posteriori*, and look at the limits along various subsequences of the data on 2 million patients. Checking whether there are any patterns in the data may be useful, but this does not change in any way the fact that making a prediction requires the ability to recognize symmetries.

If we base probability theory on the concept of a collective, we will have to apply knowledge acquired by examining one collective to some other collective. A possible way would be to combine the two collectives into one sequence and check if it is a collective. This theoretical possibility can be implemented in practice in two ways. First, one could apply a series of tests to see if the combined sequence is a collective—this would be a solid but highly impractical approach because of its high cost in terms of labor. The other possibility is to decide that the combined sequence is a collective (an exchangeable sequence) using (L4), that is, to recognize the invariance of the combined sequence under permutations. This is a cost-efficient method, but since it is based on (L4), it makes the concept of a collective redundant.

5.4 Frequency theory and the Law of Large Numbers

It is clear that many people think that a philosophy of probability called the "frequency theory" is just a philosophical representation of the mathematical theorem and empirical fact known as the "Law of Large Numbers." Paradoxically, the philosophical approach to probability chosen by von Mises makes it practically impossible to apply the Law of Large Numbers in real life.

A simple version of the Law of Large Numbers says that if we have a sequence of independent experiments such that each one can result in a "success" or "failure," and the probability of success on each trial is equal to the same number p then the proportion of successes in a long sequence of such trials will be close to p. Probabilists call such a sequence "Bernoulli trials." In von Mises' philosophical theory, probabilities are not assigned to individual events. Hence, to give meaning to the Law of Large Numbers, we must represent the "probability of success on the k-th trial" as a long-run frequency. If we want to apply the Law of Large Numbers in the frequency theory context, we must consider a long sequence of long sequences. The constituent sequences would represent individual trials in the Bernoulli sequence.

To apply the Law of Large Numbers in real life, you have to

recognize an i.i.d. sequence and then apply the Law of Large Numbers to make a prediction. The von Mises theory says that you have to recognize a collective, that is, a sequence that satisfies the Law of Large Numbers. The frequency theory failed to recognize the real strength of the Law of Large Numbers—one can use the Law of Large Numbers to make useful predictions starting from simple assumptions and observations.

Currently, it is common to limit the identification of frequency and probability to long or infinite (in theory) sequences of i.i.d. (independent identically distributed) events. The frequency interpretation applies to a much larger class of situations. First of all, the events do not have to be ordered as a sequence. The frequency interpretation applies equally well to unordered populations. Second, the additivity of frequencies holds exactly, not only approximately, for finite, not necessarily large or infinite, populations and sequences. Third, the additivity does not depend on any symmetries, independence, or other probabilistic properties of the family of events. While the first impression might be that all of these properties make the frequency interpretation an excellent theory of probability, the idea's incredibly broad spectrum of applicability exposes its weaknesses. Frequency is a useful concept in much more limited circumstances. To see this, note that the frequency interpretation applies to a completely arbitrary collection of events, related or unrelated in any way.

5.5 Computer simulations

It is quite often that we cannot find a real sequence that could help us find the probability of an event by observing the relative frequency. One of my favorite examples is the probability that a given politician is going to win the elections. I do not see a natural i.i.d. sequence (or collective) into which this event would fit. Many people believe (see [35], p. 19, for example) that computer simulations provide a modern answer to this philosophical and scientific problem. I will argue that this is not the case.

Computer simulations are an excellent scientific tool, allowing scientists to calculate probabilities with great accuracy and reliability in many cases. But they cannot replace a nonexistent real sequence. On the philosophical side, computer simulations contribute very little to the discussion.

Von Mises' theory does not support computer simulations. Natural forest fires form a collective, and computer simulations of forest fires form a collective. The two collectives are unrelated physically. To identify them as a single collective would require an extra philosophical principle. Strictly speaking, von Mises' theory does not explicitly state that the elements of a collective have to be physically "identical," but there is no explicit discussion of how to combine physically unrelated events into a single collective. The silence on this point indicates that von Mises did not approve combining physically unrelated events into a single collective.

Contemporary computers can simulate very complex random systems. Every year the speed and memory size of computers increase substantially. However, computer simulations generate only an estimate of probability or expectation that, under ideal circumstances, could be obtained in the exact form using pen and paper. In this way, computer simulations play the same role as numerical calculations (that is, deterministic computations generating an estimate of a mathematical quantity).

An event may belong to more than one "natural" sequence. One could simulate all these sequences and obtain significantly different estimates of the probability of the event. The philosophical and practical problem is determining which answers are relevant, and simulations offer no answer to this question.

Computer simulations will not turn global warming into a problem well placed in the framework of von Mises' frequency theory. The problem is not the lack of data. According to von Mises, a single event does not have a probability. No matter how many atoms you simulate, you cannot determine whether an atom is aggressive. This is because the concept of aggression does not apply to atoms. No matter how many climate change events you simulate, you cannot determine the probability of climate change event in the next 50 years. This is because the concept of probability does not apply to individual events, according to von Mises.

Chapter 6

Subjective Philosophy of Probability

Unlike von Mises, de Finetti contributed two major ideas to philosophy. The first of them, the Dutch book argument, is interesting but useless. The second one, exchangeability, is both interesting and useful. De Finetti packaged the two ideas in a philosophical theory that is a complete intellectual failure.

Recall from Sec. 4.1.4 that it is perfectly acceptable for a philosopher to deny the existence of the central object of an ideology (religion, political ideology, science), but it would be silly to deny the existence of the ideology itself. So, the philosopher faces the task of explaining social behavior wrapped around a seemingly non-existent entity.

De Finetti famously claimed that "Probability does not exist." But probability is used by mathematicians, statisticians, scientists, engineers, doctors, economists, business people, military people, and, last but not least, everybody else. One needs an explanation for the popularity of probability.

The practical applications of probability, the general field of philosophy of probability and the specific philosophical theory invented by de Finetti can be split into two major components:

(i) The mathematical rules of probability, also known as consistency or coherence.
(ii) Symmetry, independence, and all other concepts used to match real events and mathematical formulas.

De Finetti's major contribution to (i), the Dutch book

argument, is a beloved philosophical curio for some people. However, the argument is disconnected from the practical applications of probability. It is also a late addition to the list of philosophical arguments supporting (i) and the most complicated of them.

Consistency was never the subject of philosophical controversy. Justifying consistency is like justifying the claim that $2 + 2 = 4$. In principle, the claim has to be proved, but only a handful of people care about the proof.

De Finetti had absolutely nothing interesting or significant to say about (ii), so he argued that (ii) is not a problem at all. I will argue that this was a case of hypocrisy++.

6.1 Consistency

The concept of consistency (not to be mistaken with logical consistency), also known as coherence, is very close to what mathematicians call "Kolmogorov's axioms." These specify the rules that probabilities must satisfy. In Chap. 9, I incorporate them into my own theory as (L1), (L2) and (L6):

(L1) Probabilities are numbers between 0 and 1 (including 0 and 1), assigned to events whose outcome may be unknown.

(L2) If events A and B cannot happen at the same time, then the probability that A or B will occur is the sum of the probabilities of the individual events, that is, $P(A \text{ or } B) = P(A)+P(B)$.

(L6) An event has probability 0 if and only if it cannot occur. An event has probability 1 if and only if it must occur.

Kolmogorov's axioms differ from the above in two ways. First, the axioms assert that the probability measure of the whole space of outcomes is equal to 1, without a reference to reality. In general, Kolmogorov's axioms are a purely mathematical system, so they do not include any statement on the relationship between the probability values and the physical universe. Second, Kolmogorov's axioms say that probability is countably additive, that is, if $A_1, A_2, \ldots$ is a countably infinite sequence of events such that no two events can happen at the same time, the probability that at least one of these events will happen is the sum of the probabilities of the individual events. Countable additivity is irrelevant to my arguments, so I will ignore the difference between finitely additive and countably additive probability in this book.

Chronologically, the Dutch book argument is the third among the well-known philosophical theories (implicitly) justifying consistency. The earlier two are "classical" and "frequency." The classical definition of probability says that the probability of an event is the ratio of the number of cases favorable to the event and the total number of cases (see Sec. 2.1). Consistency, as stated above in the form of (L1), (L2) and (L6), immediately follows from this definition. The earliest applications of the classical definition go back at least to the mid-seventeenth century, so consistency was implicit in probability applications for several centuries. In the middle of the nineteenth century, John Venn and other philosophers developed a theory identifying probability with relative frequency (see Sec. 2.5). The laws (L1), (L2) and (L6) follow easily from the frequency interpretation of probability.

6.2 Dutch book

We first examined the Dutch book argument in Sec. 2.4. The argument justifies consistency in the form (L1), (L2) and (L6). Since the Dutch book argument is the third, in chronological order, well-known model justifying consistency, why is it popular among many philosophers and statisticians? The Dutch book argument requires fewer implicit philosophical assumptions than the preceding two. It also explicates the deterministic benefit of using probability.

The classical definition of probability can be applied only if we can reliably recognize symmetric events, i.e., events with equal probabilities. Such an ability was a subject of significant philosophical discussions and cannot be taken for granted. There is no explicit or implicit promise of any benefits of probability built into the classical definition.

Although the frequency definition of probability can be applied to almost any data, it is useful only in special cases; the i.i.d. case is best known. Hence, the frequency definition implicitly assumes that we can recognize not only "identical," i.e., symmetric events, but also independent events. In some versions of this philosophical approach, the definition implies an explicit promise of benefits. Namely, if you know the probability of an event, you can expect that the long-run frequency of the event will be very close to its probability. But this prediction is far from perfect—the frequency and probability will be close only "with very high probability," which seems to lead to a vicious circle of ideas.

In contrast to the two earlier approaches to consistency, the Dutch book argument requires only the decision-maker to recognize possible and mutually exclusive events. This is much less controversial than the implicit assumptions of the other two approaches to consistency. On top of that, the benefits of the decision-making strategy related to the Dutch book argument are deterministic. Followers of de Finetti's strategy are assured that a Dutch book will not be created against them. Finally, the predictions of de Finetti's theory can be easily tested empirically. The results of such a test would show unequivocally that the theory delivers what it promises. This contrasts with the other two definitions leading to consistency, where one would typically try to match observations and theoretical probabilities, necessitating statistical analysis.

I have pointed out in Sec. 5.4 that the frequency interpretation of probability works too well to be useful. Its scope has to be limited to be practical. A similar remark applies to the Dutch book argument. The argument is so solid that one has to suspect that there must be a catch. I will show that the argument is no more than a philosophical toy completely detached from reality. While the idea of deriving the calculus of probability from assumptions about rational decision making is intriguing and elegant, it is an empty shell.

There are three fundamental problems with the Dutch book

argument:

(i) The mathematical theorem proved by de Finetti constructs a normalized measure, not a probability measure.
(ii) There are two ways to eliminate Dutch books—the deterministic method and the probabilistic method. Do note that the latter is not unique. The preference for the probabilistic method is questionable.
(iii) Dutch books hardly ever happen. Eliminating Dutch books is as important as eliminating zombies.

Each of the three failures (i)-(iii) of the Dutch book argument has significant practical consequences and would be sufficient to disqualify the argument on its own. I will discuss (i)-(iii) in detail in Sec. 6.2.2–6.2.4.

A "Dutch book" is commonly mistaken for "any loss." De Finetti showed that probability could protect the decision-maker from a specific type of loss—a Dutch book. He did not show that probability can protect decision-makers from any other type of loss.

6.2.1 *A clarification of the Dutch book argument*

The reader may be surprised by what I plan to do in this section—to clarify and correct the Dutch book argument. This is not due to my generosity towards my intellectual opponents. I prefer "steelmanning" the problem rather than attack its over-simplified and deformed version. Consider the following objections to the rationality of the Dutch book argument.

(a) Suppose that you can make several bets whose gains or losses depend on some random events. According to the traditional definition, a Dutch book has been created against you if you lose money, no matter what outcome of some future events. De Finetti's theorem shows that if you choose any probability distribution, use it to compute the expected value of the gain of every bet, and place those bets that have positive expected gains, then you will eliminate all potential Dutch books. There is something obviously wrong with this argument—the easiest way to eliminate the potential for a Dutch book is to place no bets at all. You do not need probability to eliminate Dutch books.

(b) Suppose you can make any number of the following two bets. The first bet pays \$10.00 if the event A occurs, and it pays \$5.00 otherwise. The second bet pays \$5.00 if the event A occurs, and it pays \$10.00 otherwise. No, there is no mistake here—both bets pay a positive amount of money no matter what. You will not find yourself in a Dutch book situation no matter how many bets you make—none, one (any one) or both. But it would be irrational not to make both bets and reject an opportunity to gain \$15.00 with certainty.

(c) Suppose that you have to choose between two decisions. If you make the first decision, you will lose \$10.00. If you make the second decision, you will gain \$1.00 with a probability of 10% or lose \$100.00 with a probability of 90% (according to your subjective probabilities). The standard subjectivist advice is to make the first decision to minimize the expected loss. But the first decision will put you in a Dutch book. Were not Dutch books supposed to be eliminated by de Finetti's approach? This problem is inspired by a situation in which an investor has a choice between selling some shares of a stock today and taking a sure loss and keeping the shares but expecting that the price is likely to go down significantly.

If we modify the definition of a Dutch book in the following way, all of the above problems disappear.

Definition: Suppose that a collection of bets is available to you, and you can make any of these bets independently of your decision concerning any other bets. A Dutch book is a collection of bets $\mathcal{C}_1$, which is strictly suboptimal. That is, there exists another collection of bets $\mathcal{C}_2$ such that for every possible outcome, the combined payoff of all bets in $\mathcal{C}_2$ is greater than or equal to the combined payoff of all bets in $\mathcal{C}_1$, and there exists at least one possible outcome such that the combined payoff of all bets in $\mathcal{C}_2$ is strictly greater than that of all bets in $\mathcal{C}_1$.

Here is proof that de Finetti's strategy eliminates (redefined) Dutch books. Suppose that de Finetti's strategy resulted in $\mathcal{C}_1$ that is strictly worse than $\mathcal{C}_2$. Then, the expected gain of $\mathcal{C}_2$ is strictly greater than that of $\mathcal{C}_1$. This implies at least one of the following. Either at least one bet in the collection $\mathcal{C}_1$ was missing from $\mathcal{C}_2$ and had a negative expected gain. According to de Finetti's strategy, such a bet should not have been made. Or at least one bet in the collection $\mathcal{C}_2$ was missing from $\mathcal{C}_1$ and had a positive expected gain. This bet should have been made. Hence, de Finetti's strategy could not have generated $\mathcal{C}_1$.

Recall the three problems (a)-(c) with the old Dutch book definition. With the new definition of a Dutch book, it is not evident that placing no bets is a good strategy—this has to be verified. Hence, the decision problem cannot be trivially solved by rejecting all bets—we have a solution to (a). The new definition clearly solves problem (b). As for problem (c), losing money with probability one is not necessarily a Dutch book.

The new definition of a Dutch book disarms the main argument made in Sec. 5.5 of [2] because the argument was based on the popular version of a Dutch book. But the problem does not go away. It reappears as paranoid decision making, see Sec. 14.5.9.

6.2.1.1 *Cromwell's rule*

"Cromwell's rule," proposed in [7, Sec. 6.8], says that no event should be given probability 0 or 1. My new definition of a Dutch book gives support to Cromwell's principle.

Consider an experiment with three possible mutually exclusive outcomes: A, B and C. Suppose that you must place one and only one of the following bets at the same cost to you.

Bet I will pay \$5 in case A, −\$6 in case B and −\$7 in case C. Bet II will pay \$5 in case A, −\$6 in case B and −\$10 in case C.

It is obvious that one should place bet I and not bet II.

It is consistent to use the following probabilities: $P(A) = 0.4, P(B) = 0.6, P(C) = 0$. This coherent distribution makes bets I and II equally rational. Bet I is better than bet II, according to my definition of a Dutch book. If we follow Cromwell's rule and choose $P(C) > 0$ (and adjust the other probabilities) then placing bet I becomes the only rational strategy according to the standard decision theory. The example shows that Cromwell's rule can eliminate a decision strategy (bet II) that would also be eliminated by my version of the Dutch book argument.

6.2.2 *Normalized measures*

The standard presentation of de Finetti's theory is based on language manipulation. De Finetti proved that one could eliminate Dutch books using a normalized measure. It is customary to say that he proved that one should use "subjective" or "personal" probability.

Mathematicians and scientists borrow words from everyday language to label scientific concepts. A good example is "work," which has a considerably different meaning in physics compared to the everyday language. In physics, work is the product of distance and force. A construction worker is likely to do much more physics-type work in a day than a teacher, but many people would agree that the two people do comparable amounts of work in the ordinary sense.

A normalized measure is a mathematical function P assigning values to sets A, and such that $0 \leq P(A) \leq 1$ for all A; $P(\Omega) = 1$, where Ω denotes the whole space; and $P(A \cup B) = P(A) + P(B)$ if the sets A and B are disjoint.

A normalized measure is called "probability" or "probability measure" in mathematics. This mathematical label does not refer to any actual probability, just like work in physics does not refer to work done by a human being.

A good example of a normalized measure in real life is any representation of an object as a sum of its parts expressed in percentages. The air in Earth's atmosphere is made up of approximately 78 percent nitrogen and 21 percent oxygen. Jane ate 25% of the pie, and Mary ate 75% of the pie. California has approximately 12% of representatives in the US Congress.

So yes, it is correct to say that De Finetti proved that one could eliminate Dutch books using probability but only if "probability" represents the mathematical concept. I guess that very few people noticed this audacious language manipulation.

De Finetti's theorem proves that one can eliminate Dutch books using one's personal probabilities, one's friends' probabilities, one's financial adviser's probabilities, the composition of Earth's atmosphere in terms of percentages, the percentage distribution of the pie among the Joneses, the distribution of seats in the US Congress, or a probability distribution randomly generated by your computer. Realizing that de Finetti's theory works equally well with all normalized measures creates enormous opportunities. However, these opportunities are ignored because nobody has ever taken de Finetti's theory seriously in practical life.

Let us change the name of the measure constructed by de Finetti from "subjective probability" to "Decision Optimizing MEasure" (DOME) to see how a simple change in terminology makes a huge difference in the value and meaning of the theory. This is so as:

(i) DOME would work in a completely deterministic universe.
(ii) DOME would work in a universe governed by an evil demiurge changing the laws of nature (including the laws of probability) unpredictably.
(iii) DOME works in our universe whether objective probability exists or not. DOME works in our universe whether you have subjective probabilistic opinions or not.

(iv) Nothing in de Finetti's theory says that DOME should be equal to the objective or subjective probability if one of them exists.
(v) It is perfectly OK to delegate the choice of DOME values to others.

De Finetti's theory predicts deterministically and verifiably that any and every DOME will eliminate all Dutch books.

A term adopted in one area of science does not have to be confusing when used in a different field of science. According to [36], "Brownian motion [...] is the random motion of particles suspended in a medium (a liquid or a gas)." Another article ([37]) says, "[financial] assets have continuous prices evolving continuously in time and are driven by Brownian motion processes." It is clear that "Brownian motion" in theoretical finance does not refer to the motion of small physical particles in a liquid. The two fields of science are so far away from each other that there is no danger of any confusion. The mathematical model is the only common research element in both fields. The decision theory is intertwined with the probability theory to the point that it is very easy to miss the fact that de Finetti's normalized measures have nothing in common with the physical probability except for the mathematical structure.

6.2.3 *Who needs probability?*

The standard presentation of de Finetti's theory has a form similar to logical implication, i.e., a unidirectional argument:

$$\text{Dutch books should be eliminated} \Rightarrow \text{Probability should be used.}$$

In fact, de Finetti's theory resembles logical equivalence. If you want to eliminate Dutch books, you can use two different methods to achieve this goal:

$$\text{Deterministic verification} \equiv \text{Probabilistic decision making.}$$

Every example of a Dutch book in a philosophical or scientific book that I know of explicitly shows all negative combined payoff values of a family of bets placed using inconsistent probabilities. Hence, every example shows that one can check whether a family of bets forms a Dutch book without using probability, at least in a simple situation. On the theoretical side, one does not need probability to check whether a given decision strategy leads to a Dutch book. All one needs to do is check the bets' payoff value for every possible outcome of random events.

As far as I can tell, nobody ever checked whether the deterministic verification and elimination of Dutch books is better than the probabilistic approach in real life. The deterministic approach may be faster, cheaper, more accurate, etc. Nobody ever compared the two methods in practice. Much more is true—no philosopher or scientist ever suggested that the two methods should be compared.

Suppose two chemical research institutes, one in France and another in Germany, developed two new methods of manufacturing sulfuric acid, the world's most commonly produced industrial chemical. Chemical companies would compare the two methods before adopting any one of them. The manufacturers want to know which method is cheaper, faster, safer, etc. Nobody ever compared the quality of the two methods of eliminating Dutch books in real life.

6.2.4 *Deafening silence*

According to the World Health Organization, there were 229 million malaria cases, resulting in 409,000 deaths in 2019.

According to the Centers for Disease Control and Prevention, in 2017, 194,377 babies were born to women aged 15–19, for a birth rate of 18.8 per 1,000 women in this age group.

Magazine articles reporting on social or health maladies often supplement dry statistics with accounts of individual cases. They may devote a paragraph or two to a specific person who died of malaria or a specific teenage girl who gave birth to a child. The names of these people may be altered for privacy, but some details of their lives would be provided to stress the human dimension of the problem.

I am unaware of a single published account of an actual Dutch book. I do not mean artificial examples, such as the one in Sec. 2.4. I have never seen any statistics on the prevalence of Dutch books.

I estimate that the amount of money spent worldwide on probability is on the order of billions of dollars per year. Here is a very rough calculation. Suppose that a statistician earns about one hundred thousand dollars per year. Assuming that there are about 10 statisticians in each of the top 100 universities in the developed countries (I restrict the count to developed countries to make the salary level realistic), this brings the total expenditure on statisticians' salaries to 100 million dollars per year. One has to add to that a part of salaries paid to all other people devoting at least some of their time to probability, such as scientists, economists, doctors, business people, military personnel, etc. Next, add expenditures on hardware and software used for probabilistic and statistical calculations. Should I mention the bills for electricity and heating of the buildings? According to de Finetti, the only rational justification for all these expenditures is the desire to eliminate Dutch books.

If a politician proposes spending one million dollars on malaria prevention, wouldn't you like to know how many people are affected? If a politician proposes to spend one million dollars on teenage pregnancy prevention, wouldn't you like to know how many teenagers get pregnant every year?

Why is it that nobody ever demanded to see an estimate of the economic damage done by Dutch books every year?

De Finetti's theory is a perfectly good scientific theory. It shows how Dutch books can be eliminated using probability calculus. This theory may be useful on a different planet, where the local civilization generates many opportunities for Dutch books and where the deterministic method of eliminating Dutch books is less practical than the probabilistic one. On this planet, de

Finetti's theory is a philosophical fantasy. It is as useful as a scientific theory showing how to convert spacecraft into cheesecakes.

6.2.5 *Isolated decisions*

In classical and quantum theoretical physics, everything interacts with everything else in principle. On the other hand, in applied science and ordinary life, the world is divided into isolated systems, and only these isolated systems are analyzed. The same applies to decision making. Many (most?) decision problems are considered in isolation. The ability to isolate decision problems from the stream of life is an example of resonance (see Sec. 7.2). The separation of decisions from each other eliminates any potential for a Dutch book in many (most?) practical situations.

6.3 De Finetti's brand of skepticism

This section is analogous to Sec. 5.1.1 in that I will argue that my extreme interpretation of de Finetti's theory is the only interpretation compatible with the most prominent technical part of that theory—the decision theoretic foundations.

An axiom in the mathematical theory of probability says that if we have two events A and B that cannot occur at the same time and C is the event that either A or B occurs, then the probability of C is equal to the sum of the probabilities of A and B. In symbols, the axiom says that $P(A \cup B) = P(A) + P(B)$ if $P(A \cap B) = \emptyset$. The Central Limit Theorem, a much more complicated mathematical assertion is another example of a probabilistic statement (see Sec. 3.1.1).

A standard way to verify scientific statements such as $F = ma$, one of Newton's laws of motion, is to measure all quantities involved in the statement and check whether the results of the measurements satisfy the mathematical formula. In the case of Newton's law, we would measure the force F, mass m and acceleration a. It is essential to realize that it is impossible to check the law $F = ma$ in all instances. For example, we cannot measure any physical quantities characterizing falling rocks on planets outside of our solar system at this time. Moreover, scientific measurements have never been perfect, and perfection has never been a realistic goal, even before quantum mechanics made the perfect measurement theoretically unattainable.

If the statements that $P(A \cup B) = P(A) + P(B)$ for mutually exclusive events A and B, and the Central Limit Theorem, are scientific laws then the course of action is clear—we should measure probability values in the most objective and accurate way in as many cases as is practical, and we should check whether the values satisfy these mathematical formulas, at least in an approximate way. It has been pointed out long time ago by David Hume that, on the philosophical side, this procedure cannot be considered a conclusive proof of a scientific statement. However, verifying probability statements in this way would put them in the same league as other scientific statements.

Instead of following the simple and intuitive approach described above, de Finetti proposed to derive probability laws from postulates representing rational decision making (see p. 87 of [4] for the Dutch book argument). He proposed to limit rational decision strategies to a family nowadays referred to as "consistent" or "coherent" strategies. Then, he showed that choosing a consistent decision strategy is mathematically equivalent to choosing a probability distribution to describe the (unknown) outcomes of random events. This seems to be an incredibly roundabout way to verify the laws of probability. Would anyone care to derive Newton's laws of motion from axioms describing rational decision making? There is only one conceivable reason why de Finetti chose to derive probability laws from decision-theoretic postulates—he did not believe that one could measure probabilities fairly objectively and reliably. This agrees perfectly with his famous (infamous?) claim that "Probability does not exist." His claim is not just a rhetorical slogan—it is the essence of his philosophy.

De Finetti made a fascinating discovery that probability calculus can be used to make desirable deterministic predictions. He also declared that objective probability does not exist. The two philosophical claims are logically independent, but the first one is truly significant only in the presence of the second one.

6.3.1 *Love and hate relationship*

It is clear from the way de Finetti presented his theory that he had a love-and-hate relationship with objective probability. On the one hand, it appears that he claimed that the choice of probability values should be based on the available information about the real world, and on the other hand, he vehemently denied that probability values can be chosen or verified using objective evidence, such as symmetry or frequency.

On p. 111, Chapter II, of his essay in [38], he approvingly writes about two standard ways of assigning probabilities—using symmetry and using long-run frequencies. However, he does not invest these methods with any objective meaning. In this way, he is absolved from any philosophical responsibility to justify them. This magical trick can provide philosophical foundations for any theory, for example, the theory of gravitation. On the one hand, one can take a vaguely positive view towards Einstein's equations for gravitation. At the same time, one can declare that the available evidence of gravitation's existence is not sufficiently objective and scientific.

De Finetti has this to say about symmetry (p. 7 of [4]):

> [...] let us denote by O statements often made by *objectivists*, and by S those which a *subjectivist* (or, anyway, *this author*) would reply.
>
> O: Two events of the same type in identical conditions for all the relevant circumstances are 'identical' and, therefore, necessarily have the same probability.
>
> S: Two distinct events are always different, by virtue of an infinite number of circumstances (otherwise how would it be possible to distinguish them?!). They are equally probable (for an individual) if—and so far as—he judges them as such (possibly by judging the differences to be irrelevant in the sense that they do not influence his judgment).

On the philosophical side, de Finetti dismissed any connection between objective symmetry and probability. On the scientific side, he discussed fair coin tosses in Chapter 7 of [39]. De Finetti's book is like a treatise on tuberculosis, in which the author asserts early in the book that eating garlic has no effect on tuberculosis, but later writes several chapters on growing, storing and cooking garlic.

De Finetti defined "previsions" as follows ([4], p. 72):

> *Prevision*, in the sense we give to the term and approve of (judging it to be something serious, well founded and necessary, in contrast to prediction), consists in considering, after careful reflection, all the possible alternatives, in order to distribute among them, in the way which will appear most appropriate, one's own expectations, one's own sensations of probability.

Later in the book, de Finetti makes these remarks on the relationship between "previsions" and observed frequencies ([4], p. 207):

> *Previsions are not predictions*, and so there is no point in comparing the previsions with the results in order to discuss whether the former have been 'confirmed' or 'contradicted', as if it made sense, being 'wise after the event', to ask whether they were 'right' or 'wrong'. For frequencies, as for everything else, it is a question of prevision not prediction. It is a question of previsions made in the light of a given state of information; these cannot be judged in the light of one's 'wisdom after the event', when the state of information is a different one (indeed, for the given prevision, the latter information is complete: the uncertainty, the evaluation of which was the subject under discussion, no longer exists). Only if one came to realize that there were inadequacies in the analysis and use of the original state of information, which one should have been aware of at that time (like errors in calculation, oversights which one noticed soon after, etc.), would it be permissible to talk of 'mistakes' in making a prevision.

De Finetti's "prevision" is a family of probabilities assigned to all possible events, that is, a (prior) probability distribution. De Finetti claimed that the prior distribution cannot be falsified by the data, and neither can it be confirmed by the data. Hence, according to de Finetti, the scientific (practical) success of the theories of Markov processes and stationary processes, presented in Chapter 9 of [39], was never "confirmed" by any observations. Just like von Mises (see Sec. 5.1.2), de Finetti was (logically) inconsistent in the sense that he endorsed models for which, he claimed, there was no empirical support.

I will now go back to the question of my interpretation of de Finetti's theory. De Finetti made some conspicuous and bold statements; for example, he fully capitalized and displayed at the center of the page the claim that "Probability does not exist" (p. x of [4]). He also called this claim "genuine" in the same paragraph. On the other hand, he used the language that I find disturbingly ambivalent and confusing. For example, in the dialogue between "O" and "S" quoted above, S seems to reject in the first sentence the idea that there exist objective symmetries that would necessitate assigning the same probability value to two events. However, in the second sentence, de Finetti uses the phrase "he judges" in reference to "events ... equally probable." Very few people would interpret the phrase "he judges" as saying that an individual can arbitrarily assign probabilities, as long as they satisfy Kolmogorov's axioms. So de Finetti seems to suggest that probabilities should be chosen using some information about the real world. The following quotes confirm this opinion. De Finetti says that subjectivists follow an approach (quoted after [8], p. 218):

> leaving to a second (extra-mathematical) phase the discussion and the analysis of motives and criteria for the choice of a particular one amongst all these possible evaluations [consistent probability distributions].

Here is another de Finetti quote (after [8], p. 219):

> as for myself, though maintaining the subjectivist idea that no fact can prove or disprove belief, I find no difficulty in admitting that any form of comparison between probability evaluations (of myself, of other people) and actual events maybe an element influencing my further judgment, of the same status as any other kind of information.

I find this claim of de Finetti incomprehensible. If no facts can disprove a belief, what benefit will I derive from comparing my beliefs with others?

I have interpreted de Finetti's theory as saying that probabilities can be chosen arbitrarily as long as they satisfy the usual mathematical formulas. I consider this the only interpretation consistent with the most significant elements of de Finetti's theory. Looking into the past, he asserted that there is no objective symmetry that can be used to find probabilities. Looking into the future, he claimed that observed frequencies could not falsify any prevision. Hence, a subjective probability distribution may appear rational and anchored in reality to its holder. Still, no other person can prove or disprove objectively that the distribution is correct. For all practical purposes, this is the same as saying that probability distributions can be chosen arbitrarily.

I have already commented on logical inconsistencies in the dialogue between O and S. The quote on the frequencies is similarly annoying because it contains statements that nullify the original assertion. Suppose that at the end of clinical trials, a subjectivist statistician concludes that the side effect probability for a given drug is 2%. Suppose further that when the drug is sold to a large number of patients in the general population, the observed side effect rate is 17%. It would be very natural for the statistician to review the original study. Suppose that he concludes that the original estimate of 2% did not match the 17% rate in the general population because the patients in the clinical trials were unusually young. This seems to fit perfectly into the category of "oversights" that de Finetti mentions parenthetically. Hence, de Finetti's supporters can claim that his theory agrees well with scientific practice. De Finetti labeled as "oversight" every case where the data falsified the prevision. In this way, he did not have to discuss the philosophical problem of how observed frequencies falsify probability statements—something to which Popper devoted many pages in [14].

The strongest evidence against de Finetti is his silence. No, a philosopher cannot "take the Fifth" (exercise the right, guaranteed by the Fifth Amendment to the US Constitution, of refusing to answer questions to avoid incriminating oneself). De Finetti did not provide any criterion for objectively discriminating between different consistent probability distributions. The silence on this problem makes his theory an empty shell.

6.3.2 *Positive and negative subjectivity*

The concept of "subjectivity" has many meanings. Let us consider two of those and call them "positive" and "negative."

The positive meaning of subjectivity is that it gives freedom to people. People may choose what they eat and how they dress—"subjectivity" is a promise that other people will respect their choices and there will be no negative consequences. Negative subjectivity is the undesirable human component in accounts of events and phenomena. People try to eliminate this type of subjectivity as much as possible, especially in science.

De Finetti's philosophical theory is an example of positive subjectivity because it gives freedom to probability assessors in quite explicit ways. Bayesian statistics is an example of negative subjectivity because it tries to eliminate human bias, even if it does so implicitly and inconspicuously most of the time.

6.4 Popular subjectivism

In some cases, professional and popular versions of a philosophical theory may be considerably different. This is why a radical or impractical philosophical theory may gather popular support. Here are two examples unrelated to probability. "Immaculate conception" is the doctrine that the Virgin Mary was conceived free from original sin in the Catholic (professional) theology (see [40]). The term "immaculate conception" refers to the virgin birth of Jesus in the general culture. "Marxist rebels" in underdeveloped countries take money from the rich and give it to the poor. This is very different from Marx's (professional) theory of revolutions. Communist revolutions are supposed to happen in the most developed capitalist countries—they are supposed to install a new ruling class (the working class).

I am convinced that 99% of the support for the subjective

interpretation of probability stems from the confusion concerning the meaning of the word "subjective" and the misinterpretation of the Dutch book argument. I believe that most people subscribe to the following false interpretations of the subjective theory.

(1) Probability is subjective because different people have different knowledge bases.
(2) In practical situations, we often lack information or theories that would yield precise probability values, so we have to content ourselves with approximate values based on our intuition.
(3) The combination of the Bayes theorem, subjective probability and decision theory minimizes the expected loss.
(4) The idea of consistent probabilities is similar to Newton's laws of motion—in each case, one needs to substitute values representing objective physical quantities into the formulas.
(5) Since both the subjective philosophy and exchangeability were invented by de Finetti, they must belong to a single coherent theory.

I will show that each of the above interpretations is incompatible with de Finetti's philosophical theory.

I will first debunk (1). It is easy to see that for some events A, B and C, the conditional probability of A given B is different from the conditional probability of A given C. Formally,

$$P(A \mid B) \neq P(A \mid C) \tag{6.1}$$

for some A, B and C. Informally speaking, different people may estimate the probability of an event A in different ways because they have different prior information represented in the formula by B and C. Formula (6.1) is a part of the standard mathematical theory of probability based on Kolmogorov's axioms. It is not controversial since virtually all scientists and philosophers accept this mathematical theory. If believing in (6.1) means that one is a subjectivist, then we are all subjectivists—not a very interesting philosophical insight. It is absurd to think that de Finetti invented (6.1) or that (6.1) is the core of his philosophical theory. That would make his theory much worse than wrong—it would make it trivial. The fundamental claim of de Finetti's philosophical theory can be represented by a different formula, namely, $P_1(A \mid B) \neq P_2(A \mid B)$. De Finetti claimed that two people possessing the same information might differently assess the probability of a future event (P_1 vs. P_2), and that there is no objective way to verify who is right and who is wrong.

(2) It is obvious that, quite often, we do not have sufficient information such as facts or knowledge of reliable theories to make accurate probability assessments. It is equally obvious that, quite often, we do not have sufficient information such as facts or knowledge of reliable theories to make accurate assessments of temperature, distance, electrical charge, etc. In everyday life, we may have limited access to measuring instruments. Nobody will refrain from saying "my coffee is cold" (if she believes that this is the case) just because she has no thermometer to measure the temperature accurately. Scientists cannot measure many physical quantities with great accuracy for a whole spectrum of reasons, from profoundly theoretical (Heisenberg's uncertainty principle) to practical (some galaxies are too distant) to trivial (lack of money). I am unaware of "subjective philosophies" of temperature, distance, electrical charge, etc. Of course, psychologists study human perceptions of probability, temperature, distance, etc., but this does not indicate in any way anyone's belief that these concepts are subjective. If "subjective" means that a quantity is sometimes informally evaluated, then probability is subjective, but so are temperature and distance. De Finetti did not create a subjective philosophy for all scientific quantities. So, his idea of "subjective probability" must have differed from the informal objective probability assessment.

(3) De Finetti's theory shows how probability can be used to eliminate Dutch books. A Dutch book is a very specific and limited situation that leads to loss. Nothing in de Finetti's theory indicates how to avoid or minimize any other type of loss. Since

all consistent probabilistic opinions are considered equally rational (or rather, the subjective theory does not offer any method to compare their benefits), an almost unlimited family of decision strategies are "rational." Each of these strategies minimizes its expected loss. Still, there is no connection between this expected loss and reality because there is no connection between probability and reality, according to de Finetti. The subjective theory implies the paradoxical fact that incompatible (but individually consistent) strategies maximize the expected gain. This is a purely abstract claim with no practical meaning.

(4) If engineers want to send a satellite to the Moon, they have to use, among other things, Newton's "second law" saying that $F = ma$, where F denotes force, m denotes mass and a denotes acceleration. Imagine that one of the engineers proposed that the mass should be $7,777,777$ kilograms. She likes the number seven; hence, she chose seven sevens for the mass. Next, suppose another engineer proposed that the acceleration should be 10^{10}m/s^2 because this is a nice round number. Given these two numbers, F is necessarily equal to $7,777,777 \cdot 10^{10}$. My story sounds like a bad joke. Engineers never arbitrarily choose numbers, satisfying their whims. The numbers that they use must represent real quantities, in the present case, the combined mass of the rocket and the satellite, and the acceleration needed to put it in the correct orbit.

Without question, Bayesian statisticians try to match probabilities in their statistical models with reality. They treat de Finetti's consistency idea just like Newton's laws of motion—these are mathematical formulas that have to be filled with values characterizing the real universe. I am unaware of any evidence that any statistician arbitrarily assigns probabilities, as in de Finetti's theory.

(5) Exchangeability was invented and promoted by de Finetti. Exchangeability is a great contribution to mathematics—it is a very elegant and powerful purely mathematical concept. Exchangeability is also a great contribution to science—it is one of the two most important concepts underlying Bayesian statistics

(the Bayes Theorem is the other one). According to de Finetti's own theory, exchangeability is useless. He claimed that there is no objective symmetry—this includes exchangeability. He gave no reason for why using exchangeability (in the objective or subjective sense) would benefit the decision-maker. Mathematicians, statisticians, and other scientists know the significance of exchangeability in mathematics and science. They do not realize that de Finetti's philosophy strips the concept of any practical value. Many popular accounts of the subjective philosophy suggest implicitly or state explicitly that the subjective philosophy is inextricably connected with exchangeability.

6.4.1 *Textbooks*

The subjective philosophy of probability is linked to Bayesian statistics in the minds of many people. Of the two fields of intellectual activity, Bayesian statistics boasts overwhelmingly larger membership than the narrow field of the philosophy of probability. Bayesian statisticians are trained in colleges using textbooks and, at the most advanced level, research articles and monographs. Bayesian textbooks are filled with examples of Bayesian data analysis. The examples follow clear patterns; for example, many of them incorporate exchangeability. Students of Bayesian statistics might know about its link to the subjective philosophy of probability, but very few of them read books on philosophy. Hence, they assume that "subjective probability" is the label for the contents of Bayesian textbooks. This could not have been further from the truth.

6.4.2 *Spiritual needs*

I have already argued that quite strong and misguided support for the subjective philosophy among Bayesian statisticians is due to misconceptions and confusion. Misconceptions are due to ignorance. Confusion is due to contradictory claims made by subjective philosophers.

The support of the subjectivism among Bayesian statisticians and the frequency interpretation of probability among frequency (classical) statisticians have partly roots in statisticians' spiritual (emotional) needs. The two branches of statistics seem unable to resolve scientific differences between them—a rather embarrassing situation for any branch of science. The problem is not that the scientific controversy has not been resolved yet but that there are no widely agreed-upon ideas on how to resolve the conflict. So it is natural that people on both sides of the controversy seek support from a religious-like source, that is, philosophical theories harboring beliefs in conflict with the empirical reality.

6.4.3 *Seductive theory*

I do not want to leave an impression that the subjective philosophy of probability is popular exclusively due to ignorance and confusion. It has several aspects that can and do make it very attractive to some people.

(i) The subjective philosophy is based on a mathematical theorem. This does not make it unique among philosophical theories of probability because the frequency theory is related to the Law of Large Numbers (I say "related" because von Mises' version of the frequency theory is based on collectives that assume relative frequencies' stability instead of proving it).
(ii) The subjective philosophy assures its users a deterministic benefit (elimination of Dutch books), something that the frequency theory does not promise. The core of the logical philosophy of probability does not explicitly promise anything.
(iii) The subjective philosophy makes empirically verifiable predictions. It is effectively a branch of science.
(iv) In the subjective theory, probability is derived from decision-making theory rather than vice versa. The idea of reversing the relationship between concepts appeared in both

philosophy and science. In philosophy, Ludwig Feuerbach asserted that "God did not, as the Bible says, make man in His image; on the contrary man [...] made God in his image." On the mathematical side, one can prove that Brownian motion is a continuous martingale whose quadratic variation equals t (time). One can reverse the logic and define Brownian motion as unique (in the sense of distribution) continuous martingale whose quadratic variation is equal to t. One can reasonably argue that the reversal of roles of concepts adds depth to the analysis, even if it is not correct or applicable.

6.4.4 *Professional philosophers on subjectivism*

I have to admit that I have a feeling that my book may be redundant. My central claims have already been published. Consider the following three examples of opinions expressed by professional philosophers studying probability (among other things).

In [22], we find

> Orthodox Bayesians in the style of de Finetti recognize no rational constraints on subjective probabilities beyond:
>
> i. conformity to the probability calculus, and
>
> ii. a rule for updating probabilities in the face of new evidence, known as conditioning. [...]
>
> This is a permissive epistemology, licensing doxastic states that we would normally call crazy.

In Chap. 5 of [41], the author says

> The representation theorems of ordinary decision theory show that an agent's preferences are represented by *some* probability measure, but they put no constraint at all on which one. [emphasis in original]

See also Chap. 4 of [41] for a detailed discussion of some philosophical options.

According to [42, p. 171]:

> Subjectivistically interpreted, Normative Bayesian Decision Theory, whatever its heuristic virtues, is either philosophically vacuous or impotent.

In a sense, Hájek, Kyburg and Wallace said it all. However, the reader could miss the crucial point because it is not spelled clearly enough in those publications. Both [22, 41] say that to avoid "crazy doxastic states," i.e., "crazy beliefs," one has to adopt some extra assumptions. They say that these additional assumptions are commonly taken to be symmetry or identification of frequency and probability. What I am missing in [22, 41] is a clear statement of the following consequences: (i) On the practical side, the extra assumptions require a commitment identical to that made in the case of quantities considered to be objective. (ii) On the philosophical side, the additional assumptions require analysis and justification customarily reserved for "objective" claims.

My objection to Kyburg is somewhat different. He says (see [42, p. 176]),

> Given that two people have degrees of belief that satisfy the probability calculus, and given that their degrees of belief satisfy some relatively mild constraints in addition to the coherence constraints, then, as empirical evidence accumulates, their distributions of beliefs will become more and more nearly the same.

The "relatively mild constraints" would typically mean exchangeability. The assumption of exchangeability is far from "mild."

Chap. 16 is devoted to reviews of two recent books on the foundations of probability: [6, 7]. In these two books, subjectivism is presented strikingly differently than in the quotes from [22, 41, 42]. Although some professional philosophers fully grasped the essence of subjectivism, the two books contain awfully muddled accounts of subjectivism.

I do not want to leave an impression that professional philosophers are the only intellectuals capable of understanding the weaknesses of de Finetti's theory. Jim Berger is a very prominent professional statistician. He supports Bayesian methods in statistics—see [43]. The axiomatic approach to Bayesian statistics is described in Sec. 4.1. IV of [43]. Berger points out that axiomatic systems do not prove that "*any* Bayesian analysis is good."

The book [44] is devoted to carefully examining the subjectivity of many prominent scientists. The authors state on p. 202:

"Or it could be that neither definition is appropriate and the subjective probability statement instead implies a belief of the individual based on their scientific knowledge and deep understanding of some underlying biological, physical, or social phenomenon." The suggestion that a deep understanding of the physical universe is a part of subjective probabilistic beliefs is miles apart from de Finetti's philosophical theory.

6.5 The subjective theory of probability is objective

The labels "personal probability" and "subjective probability" used for de Finetti's theory are highly misleading. His theory is objective and has nothing to do with any personal or subjective opinions. De Finetti says that certain actions have a deterministic result, namely, if you take actions that are coordinated in a way called "consistent" or "coherent," then you will eliminate all Dutch books. This argument has nothing to do with the fact that the decision maker is a person. The same argument applies to other decision-makers: businesses, states, computer programs, robots and aliens living in a different galaxy.

In practice, probability values in de Finetti's theory have to be chosen by people, but this does not make his theory any more subjective than Newton's laws of motion. A person has to choose a body and force to be applied to the body, but this does not make acceleration predicted by Newton's laws subjective. Newton's claim is that acceleration depends only on the body's mass and the strength of the force and has nothing to do with the personal or any other way in which the body and force have been chosen. Similarly, de Finetti's predictions are objective. If a monkey chooses a consistent decision strategy by randomly pounding on the keyboard, de Finetti's theory makes an objective and verifiable prediction that no Dutch book will be formed against the beneficiary of monkey's decisions.

This brings us to a closely related issue that needs to be clarified. In de Finetti's theory, one should talk about a beneficiary and not a decision maker. The two can be different; for example, a computer can be the decision-maker and a human can be the beneficiary, or an employee can be the decision-maker and her employer can be the beneficiary. De Finetti's analysis makes predictions concerning the beneficiary. The decision-maker is in the background; he is almost irrelevant. The use of adjectives "personal" and "subjective" is misleading because it suggests that the decision-maker is the main protagonist of the theory. In fact, the theory is concerned exclusively with predictions concerning the beneficiary.

De Finetti is sometimes portrayed as a person who tried to smuggle a non-scientific concept (subjectivity) into science, like an alchemist or astrologer. Nothing can be further from the truth. De Finetti wanted to purge a non-scientific concept—probability—from science. De Finetti did not promote the idea that one should use personal or subjective probability because doing so would be beneficial. He was saying that choosing probabilities in a consistent way was necessary if one wanted to avoid deterministic losses. Choosing probabilities by applying personal preferences is practical, but there is nothing in de Finetti's theory that says that using any other way of choosing probabilities (say, using computer software) would be less beneficial.

A chemist or engineer could propose a "theory of painting" by claiming that "painting a wooden table increases its durability." The following practical choices do not affect this prediction: (i) the color of the paint (as long as the paint chemical composition satisfies appropriate quality standards), (ii) the gender of the painter, (iii) the day of the week. De Finetti made a scientific and verifiable prediction that you can eliminate Dutch books if you use probability. The following choices have no effect on de Finetti's prediction: (i) the probability values (as long as they satisfy Kolmogorov's axioms), (ii) the gender of the decision-maker, (iii) the day of the week. De Finetti's theory is no more subjective than the "theory of painting."

6.6 Coincidences

When we find a pattern, we try to understand it. Take, for example, the motions of celestial bodies. The motions are very regular but not perfectly so. The Ptolemaic system was the best-known ancient attempt to understand the motions of the planets, but it pointed in the wrong direction. The contribution of Copernicus was to interpret the motions in a different, heliocentric way. This led to further improvements in the theory by Kepler, who used ellipses to model the planets' trajectories. The next step was made by Newton, who invented gravitation to explain the motions of the planets and terrestrial objects. Einstein changed the view of gravitation by interpreting it as the spacetime curvature. At every step, the new theory was deeper than its predecessor because it explained the already known phenomena and provided a basis for new predictions. At every stage, science consists of some claims that have to be taken without an explanation—they only assert that some regularity has been observed. Currently, many physicists believe that string theory is the ultimate theory of everything. But even if it is true, the string theory remains without an explanation.

In his analysis of probability, de Finetti stopped at an embarrassingly shallow place. He said (quote from [9], p. 70):

> Our point of view remains in all cases the same: to show that there are rather profound psychological reasons which make the exact or approximate agreement that is observed between the opinions of different individuals very natural, but there are no reasons, rational, positive, or metaphysical, that can give this fact any meaning beyond that of a simple agreement of subjective opinions.

De Finetti's remark reveals complete intellectual failure, pretends to be profound, and contains a striking logical inconsistency.

"The exact or approximate agreement that is observed between the opinions of different individuals" is stated as a fact and left without explanation. We cannot blame Aristotle for not discerning gravitation in the motions of the planets, but at least Aristotle did not claim that "there are no reasons, rational, positive, or metaphysical, that can give this fact any meaning beyond that of a simple agreement of" the regularities in planet trajectories.

There are many examples of exact or approximate agreement between opinions outside probability. Their explanations include passing scientific or proto-scientific knowledge between people, indoctrination of children, tradition combined with social pressure, and herd instinct. De Finetti's intellectual failure is especially striking because other philosophers proposed explanations of the agreement between people on probabilities based on symmetry and frequency. All de Finetti could do was dismiss those philosophical ideas without proposing alternatives.

The initial remark about "profound psychological reasons" is fake profundity. If the reasons are indeed profound, then they should be explicated. Once again, other philosophers proposed explanations of the "profound psychological reasons" based on symmetry and frequency.

Finally, juxtaposing "profound reasons" and "simple agreement" is just a logical inconsistency.

6.7 A science without empirical content

De Finetti's theory fails one of the basic tests for a scientific theory of probability—it does not report any probabilistic facts or patterns observed in the past. I will illustrate this claim with examples from physics and probability.

Facts and patterns can be classified according to their generality. Consider the following facts and patterns.

(A1) John Brown cut a tree branch on May 17, 1963, and noticed the saw was hot when he finished the task.
(A2) Whenever a saw is used to cut wood, its temperature increases.
(A3) Friction generates heat.
(A4) Mechanical energy can be transformed into heat energy.
(A5) Energy is always conserved.

I might have skipped a few levels of generality, but I am sure the example is clear. Here are probabilistic counterparts of the above facts and patterns.

(B1) John Brown flipped a coin on May 17, 1963. It fell heads up.
(B2) About 50% of coin flips in America in 1963 resulted in heads.
(B3) Symmetries in an experiment such as coin tossing or in a piece of equipment such as a lottery machine are usually reflected by symmetries in the relative frequencies of events.
(B4) Probabilities of symmetric events, such as these in (B3), are identical.

I consider the omission of (B4) from the subjective theory to be its fatal flaw that destroys its claim to be a scientific theory representing probability. I will examine possible excuses for the omission.

Some sciences, such as paleontology, report individual facts at the same level of generality as (A1) or (B1), but I have to admit that the theory of probability cannot do that. One of the reasons is that the number of individual facts relevant to probability is so large that they cannot be reported in any usable way, and even if we could find such a way, the current technology does not provide tools to analyze all the data ever collected by humanity.

The omission of (B2) by the subjective theory is harder to understand, but it can be explained. Most people consider this type of information useful and relevant. Truly scientific examples at the level of generality of (B2) do not deal with coin tosses. Scientists are concerned with repeated measurements of scientific quantities, for example, the frequency of side effects for a drug. It is a legitimate claim that observed patterns at this level of generality belong to various fields of science such as chemistry, biology, physics, etc. They are reported by scientists working in these fields, so there is no need to incorporate them into the theory of probability. One could even say that such patterns do not belong to the probability theory because they belong to some other sciences.

Finally, we come to (B3) and (B4). These patterns do not belong to any science such as biology or chemistry. If the science of probability does not report these patterns, who will?

If you roll a die, the probability that the number of dots is less than 3 is 1/3; this is a concise summary of some observed patterns. Every theory of probability reported this finding in some way, except for the subjective theory. Needless to say, de Finetti did not omit such statements from his theory because he was not aware of them—the omission was a conscious choice. De Finetti's choice can be easily explained. If he reported any probabilistic patterns, such as the apparent stability of relative frequencies in long runs of experiments, his account would have taken the form of a "scientific law." Scientific laws need to be verified (or need to be falsifiable in Popper's version of the same idea). Stating any scientific laws of probability would have destroyed de Finetti's philosophical theory. The undeniable strength of his theory is that it effortlessly avoids the thorny question of the verifiability of probabilistic statements—it denies that there are any objectively true probabilistic statements. The same feature that is a philosophical strength is a scientific weakness. No matter how attractive the subjective theory may appear to philosophically minded people, it has nothing to offer on the scientific side.

6.8 The essence of probability

I will use the term "agent" to designate a person or a computer program. It might be easier to imagine a computer program making strange decisions, rather than a human being acting that way. Consider four agents, applying different strategies in the face of uncertainty.

(A1) Agent A1 assigns probabilities to events without using the mathematics of probability, without using consistency and without using (L1)-(L6). He does not use any other guiding principle in his choices of probability values.

(A2) Agent A2 is consistent but does not use (L1)-(L6).

(A3) Agent A3 uses (L1)-(L6) in his probability assignments but does not use the mathematical rules for manipulating probability values.

(A4) Agent A4 applies both (L1)-(L6) and the mathematical theory of probability (in particular, he is "consistent").

Let me make a digression. I guess that agent A3 is a good representation for a sizeable proportion of the human population. I believe that (L1)-(L6) are at least partly instinctive (see Sec. 7.2 on resonance), and so most people use them but the mathematical rules of probability are not easy to apply at the instinctive level, and they are mostly inaccessible to people lacking education. Whether my guess is correct, it is inessential since I will focus on agents A1, A2 and A4.

Before I compare the four agents, I want to comment on the interpretation of the laws of science. Every law contains an implicit assertion that the elements of reality not mentioned explicitly in the law do not matter. Consider the following example. One of Newton's laws of motion says that the acceleration of a body is proportional to the force acting on the body and inversely proportional to the mass of the body. An implicit message is that if the body is green and we paint it red, doing this will not change the acceleration of the body. (This interpretation is not universally accepted—some young people buy red cars and replace standard mufflers with noise-making mufflers, hoping that the red color and noise will improve the acceleration of the car.)

It is pretty clear that agents A1 and A4 lie at the two ends of the spectrum when it comes to success in ordinary life, but even more so in science. Where should we place agent A2? I do not doubt that A2 would have no more than a 1% greater success rate than A1. In other words, consistency can account for less than 1% of the overall success of probability theory. I guess A3 would be about halfway between A1 and A4, but such speculation is unnecessary for my arguments.

Now, I am ready to argue that the subjective theory of probability is false as a scientific theory. The theory claims that probability is subjective, there is no objective probability, and you must be consistent. An implicit message is that if you assign equal probabilities to symmetric events, as in (L4), you will not gain anything, just like you cannot increase the acceleration of a body by painting it red. Similarly, the subjective theory claims that using (L3) cannot improve your performance. In other words, the subjective theory asserts that agent A2 will do as well in life as agent A4. I consider this assertion absurd.

De Finetti failed spectacularly by formalizing only this part of the probabilistic methods, which explains less than 1% of the success of probability—he formalized only the consistency, that is, the necessity of applying the mathematical rules of probability.

The subjective philosophy is walking on a tightrope. It must classify some decision families or probability distributions as "rational" and some as "irrational." The best I can tell, only inconsistent families of probabilities are branded irrational. All other families are rational. Moving some consistent families of decisions, that is, some probability distributions, to the "irrational" category would destroy the beauty and simplicity of the subjective philosophy. Leaving them where they are, makes the theory decoupled from the scientific practice.

When designing a theory, one can either choose axioms that are strong and yield strong conclusions or choose axioms that are weak and yield few conclusions. There is a wide misconception among statisticians concerning the strength of de Finetti's theory. The axioms of his theory are chosen to be very weak so that they are acceptable to many people. The price you must pay for this intellectual choice is that the conclusions are incredibly weak.

Philosophers and scientists invented many things that are not observable. Plato invented "forms" also known as "ideas." Kant invented "things in themselves." Someone invented God and angels. Physicists invented superpositions of quantum states—think about Schrödinger's cat. Physicists also invented electric fields and protons that are not directly observable by people. One of the roles of unobservable things is to help summarize our knowledge. Whether Plato or Kant succeeded, I will leave it to the reader to decide. Physics is widely considered to be very successful. The probability of an event is unobservable in a single experiment. The question is whether the objective probability of a single event is valuable fiction. Everything we know about science says that it is.

6.9 Theology

De Finetti's theory is just like St. Thomas Aquinas' proofs of God's existence ([45]) or Pascal's wager ([46]). They stop exactly where things start to be interesting and practical. Saying that God exists does not say anything about how many wives a man can have at a given time: one (according to Catholics), four (according to Muslims), or an unlimited number (according to a branch of Mormons). Likewise, the issue of divorce is not resolved by St. Thomas and Pascal's arguments—whether it is permissible or not. Most people worldwide believe in God or gods. Proving to them that God exists is redundant. But in some countries, followers of different religions reside. Marriage laws are an important practical matter. The philosophical arguments of St. Thomas and Pascal spectacularly fail to give any relevant advice on marriage.

De Finetti gave an argument in support of consistency, i.e., that probabilities are between 0 and 1, and that probabilities should form a (finitely) additive measure. No mainstream philosophy had ever contested these ideas. The real question always was how to assign probability values to real-life events. De Finetti's theory has absolutely nothing to say about it.

St. Thomas's proofs of God's existence and Pascal's wager are understood in a completely different way by their proposers and followers, on the one hand, and atheists on the other hand. The proposers believed in Christian God and they were only seeking arguments to dispel their doubts (or doubts of people thinking likewise). Once they found proofs of God's existence, their faith was reassured and strengthened. They did not need a method to find the true God.

An atheist position is entirely different. Even if the proof that God exists is correct, it contains no helpful information—it is a useless piece of philosophical sophistry. Several major religions and many minor religions exist now. Many more religions existed in the past. The number of religions envisaged in the literature (fiction, science-fiction, etc.) and those that are possible but not yet brought to light is infinite. Religions differ dramatically in descriptions of God or gods, their attributes, their relationship with people, ethical systems imposed or supported by gods, etc. The arguments of St. Thomas and Pascal have absolutely nothing to say about the specific nature of God. They are meaningless to an atheist.

De Finetti's theory plays a role similar to that of the proofs of God's existence for the religious people with doubts. Bayesians have doubts about their method, especially the role of prior distributions. De Finetti's philosophy supposedly justifies the use of prior distributions. Once this justification is accepted, Bayesians do not need guidance on how to find good prior distributions.

6.10 Cohabitation with an evil demiurge

Imagine a world governed by an evil demiurge bent on confusing people and making their lives miserable. The demiurge changes the laws of nature in arbitrary and unpredictable ways. Sometimes, he announces the timing and nature of changes. Sometimes, he cheats by making false announcements; sometimes, he makes changes without announcing them. Some changes are permanent; for example, he changed the speed of light at some point. Some changes are temporary; for example, coffee was a potent poison for just one day. Some changes may affect only one person; for example, John Smith was able to see all electromagnetic waves with his own eyes for one year, but nobody else was affected similarly. The idea of a world with an evil demiurge is very similar to an idea of a chaotic and unpredictable world invented by David Hume as a part of an argument exposing a logical weakness in the principle of induction.

In a universe governed by the evil demiurge, no knowledge of the past events can be used to make a deterministic prediction of any future event. Likewise, statistical predictions do not have the same value as in our universe. For example, the demiurge once changed the properties of a new drug being developed by a pharmaceutical company. The drug had been 80% effective in medical trials, but it became 1% effective when the company started selling it.

The Dutch book argument would work in a world governed by an evil demiurge just fine because there is nothing in de Finetti's theorem that would invalidate the argument. Isn't it suspicious that, according to subjectivists, Bayesian statistics would work perfectly well in a world devoid of stable laws of nature?

6.11 No need to collect data

In many practical situations, scientists are not limited to studying existing data sets, but they can choose whether to collect some data or not. As long as they have interest in the subject and sufficient resources, such as money, workforce and laboratories, they inevitably choose to collect data. De Finetti's theory fails to explain why they do so and does so at many different levels, so to speak.

The easiest way to see that there is no need to collect data is to notice that, according to de Finetti, the only purpose of the probability theory is to eliminate inconsistent decision strategies. In every situation, no matter what information you have, there is at least one consistent strategy. Hence, if you do not collect any data, you can still act in a consistent way. Quite often, not collecting data would result in substantial financial savings, so the choice is obvious—stop collecting data.

Suppose the government imposes a new law saying that statisticians and scientists cannot collect more than 5 data for any project. How would this create Dutch books? Bayesian posterior distributions would be different under the new law from the current posterior distributions but Bayesian statistics would stay consistent.

Consider two people who plan to use a deformed coin to play some games. Suppose that one of the players tossed the coin 100 times, without showing the results to the other player. According to de Finetti, the first player has no advantage over the second player, whatever games they might play. There is no objective probability, so the first player could not have learned the actual or even approximate value of the probability of heads. Both players can be consistent.

Suppose you learn that a scientist falsified a data set. Will you be able to find inconsistencies in his or anyone else's decisions? Scientists insist on using only genuine data sets, but this is not because anyone ever found himself in a Dutch book situation as a result of someone else publishing a fake data set.

I will argue that scientific articles based on Bayesian statistics contain no useful information according to de Finetti's theory. A report on the results of Bayesian statistical analysis may contain the prior distribution, the data, and the posterior distribution, or only some of these.

The prior and posterior distributions are subjective in the sense of "arbitrary." De Finetti claims that there are no objective probability distributions so the reader can safely ignore the published prior and posterior distributions and come up with her own.

The data are not needed to develop a consistent probabilistic opinion on future events. The data can be relevant only when some decisions' payoffs are based on the data. It hardly ever (never, as far as I can tell) happens that a decision's payoff depends on the data published in a research article.

6.12 Probability and induction

Consider the following two sequences of binary observations (sequences of 0's and 1's) of some physical phenomena.

$$1111111111111111111111$$
$$010101110101000010100110$$

Can you guess the next element in each of the sequences?

Guessing that the first sequence would continue with a 1 is the essence of science. Within science, the guess would be considered trivial. The method applied by scientists is called "induction." The justification of induction has the reputation of a hopelessly intractable philosophical problem. See Sec. 7.1 for an introduction to the problem of induction.

Guessing the next element of the second sequence is the essence of statistics. The task is far from trivial, especially if no information is provided about the natural phenomenon and how the data were collected. Now comes the amazing—subjectivists believe the problem has a rational, elegant, and complete solution.

The coexistence of the "hopeless" induction problem and the "completely solved" (in the eyes of subjectivists) probability problem is one of the most curious aspects of the current epistemology.

6.13 Delegation of duties

We delegate things to do to other people. A physicist may delegate a measurement of temperature, energy or entropy to another physicist. Physicists are a community of people who can delegate measurements to each other and expect that the other party will understand the requests and provide results just like those produced by the originator.

Delegation is not limited to science. Another example is religion. A Catholic may ask someone (for example, a priest) to pray for the soul of a deceased relative.

The subjectivist claim that there are no objective symmetries collides head-on with applied statistics. Statisticians are a community of scientists who can delegate collecting and analyzing data to each other. In particular, they can recognize events that are i.i.d., exchangeable, symmetric and independent in an imperfect but reliable way.

6.14 Imperfect skepticism

De Finetti was brave enough to reject common sense, but only when it suited his intellectual interests. Otherwise, he was boringly mainstream. The Dutch book argument shows that de Finetti believed that facts (events) existed objectively and we could objectively determine their status. These two assumptions, one metaphysical and one epistemological are standard implicit assumptions in science.

The Dutch book argument refers to bets, that is, decisions with monetary payoffs. If one signs an agreement that entitles the signatory to a payoff, then this money will be paid in the case when the event occurs. This expectation is based on experience—signing a contract was followed by the payoff in the past. Hence, induction is built into the Dutch book argument.

There is also a social angle to my argument. Executing a bet payoff requires that the two parties recognize objective symmetries and agree on objective symmetries. For example, if the bet concerned rain on a specific day, the parties must recognize that the day was rainy; that is, it was like (symmetric with; identical to) other rainy days in the past. So, implicitly, they have to accept the existence of objective symmetries. The parties must agree on the objective outcome, or otherwise, there would be a conflict with no rational way to resolve it. De Finetti implicitly assumes that there are objective symmetries in our world, and we can reliably recognize them.

6.15 Is the speed of light a random variable?

Some people believe that a significant difference between Bayesian and frequency statistics is how they represent scientific constants. Suppose that a physicist plans to make repeated measurements of the speed of light. The standard frequency model would assume that the measurements are i.i.d. (independent and identically distributed), with the mean equal to the unknown *constant* θ (the speed of light). The standard Bayesian model would assume that the measurements are exchangeable. It would also involve the prior distribution of θ, thus making the speed of light implicitly a *random variable.* For example, θ could be normally distributed under the prior.

I will argue that de Finetti's theory makes it impossible to treat θ as a random variable. If θ is a random variable then the statement $\{\theta < 299,792\}$ is an event. (If the reader is puzzled by the number 299,792, I rush to explain that the current estimate of the speed of light is 299,792 km/s.) In de Finetti's theory, probability can be used only to protect the decision-maker from a Dutch book. But the event $\{\theta < 299,792\}$ cannot be used in any bet. A bet is an agreement between two parties. After observing the event that is the basis of the bet, one party will pay some money to the other one. This requires an objective determination of whether or not the event occurred. It is impossible to determine whether $\{\theta < 299,792\}$ holds because θ is unknown—the point of making measurements of the speed of light is to estimate the unknown quantity θ. One could base a bet on $\overline{X}$, the average of the measurements because it is possible to determine whether an event such as $\{\overline{X} < 299,792\}$ occurred. But this makes $\overline{X}$, not θ, a random variable.

PART 2
Epistemology and Science

Chapter 7

Epistemology

The first part of the book focused on the critique of two dominant philosophies of probability. The second part of the book will be constructive. I will present my vision of the science of probability, but I will examine some problems in (general) epistemology before I do that. A thorough discussion of the concept of probability, especially of the claim that probability is objective, necessitates a detailed analysis of the general foundations of knowledge creation. The primary purpose of the present chapter is to define and analyze a concept of "resonance."

7.1 Problem of induction

The simplest version of the problem of induction is concerned with a sequence of identical observations made in the past. Why is it rational to believe that similar events will occur in the future? For example, suppose that we observed that ripe apples fall from the tree to the ground. Why is it rational to expect that ripe apples will fall to the ground in the future? Is it possible that apples will fly to the sky instead?

A more general form of the problem of induction asks why it is rational to hold beliefs about the unknown based on the known. For example, scientists currently predict climate change (earlier referred to as global warming). This event is not believed to be identical to any event in the past. The climate did change in the past; for example, the Earth went through an ice age that ended about 12,000 years ago. However, no past climate changes occurred concurrently with human industrial activity. So why do many people believe that the climate will change in the near future?

On the philosophical side, the problem of induction stems from the limitations of the deductive logic. This type of logic can only transform a true proposition or a family of true propositions into a new true proposition. Deductive logic is traditionally and widely recognized as the most solid tool to generate reliable knowledge. One could dismiss the problem of induction as an artificial problem of our own making, created by our irrational attachment to one particular intellectual technique of attaining truth. I do not dismiss the problem of induction because rejecting deductive logic or arbitrarily supplementing it with other methods does not lead to a deeper understanding of the problem.

It is possible that no inductive logic will ever succeed for social reasons. Based on the history of the philosophical research of induction so far, I guess that there is little chance that a universal agreement of all philosophers on a single version of inductive logic will ever materialize.

The problem of induction is not only a philosophical problem—it is also a practical problem. We use information retrieved from our memory or obtained from other people. Since some of this information is false and some of it is unreliable, we want to understand how to tell the difference between reliable and unreliable mechanisms of generating knowledge. Ideally, inductive logic should be similar to deductive logic—mechanically verifiable and hence very reliable.

7.1.1 *An ill-posed problem*

I will argue that the traditional problem of induction is ill-posed. A brief summary of my argument is the following. The problem of induction is to provide a convincing justification for why we can pass from our information about the known to our beliefs about the unknown. We cannot pass from the known to the unknown because there is nothing that we really know. We "know" facts only because we make some assumptions about our universe and our knowledge. We need to adopt these same assumptions to make predictions. These assumptions cannot be proved, but most people are comfortable making them, consciously or subconsciously. The assumptions apply equally to the known and the unknown, so the problem of induction is an illusion on the philosophical side. Of course, we have to distinguish between the known and unknown in everyday life and science—I will discuss this point later.

The first of the implicit assumptions that we make when we talk about known facts is that our universe is governed by stable laws. This is a reasonable assumption in my opinion because otherwise, there is nothing interesting that we can say about anything. But the stability of our universe is nonetheless an assumption, not a fact. There are several ways in which our universe might not be governed by stable laws. It is possible that our universe is governed by a capricious demiurge who may change the laws of nature at any time. So, starting tomorrow, all physics may be wrong. It is also possible that the whole universe was created by the demiurge five thousand years ago together with humans and dinosaur fossils. The reader will recognize here the actual belief of a large religious community—the belief dismissed by science but impossible to dismiss on purely logical grounds. Modern technology inspires other scenarios. The film "Matrix" was concerned with an artificial world, very much like ours, existing in the memory of a computer. Scientists tell us that our physical universe extends enormously in space and time and speculate that there are many other universes (see [47]). This lead some people to speculate that some matter might have assembled randomly somewhere in the universe and formed a mind with the thoughts that one of us (the reader of this book?) experiences (see [48]). These thoughts do not have to have anything in common with the existing universe. The common aspect of all these scenarios is that they all allow for the universe to have been completely different from the current universe at any time in the past. It follows that the "fact" that all apples fell to the ground in the past is based on the *assumption* that the world was stable until now.

While the first assumption, discussed above and made implicitly when we assert "facts," has an ontological nature, the second one, to be discussed next, is epistemological. Let us take for granted that there exists an objective universe governed by stable laws. Do we have the ability to observe the universe and correctly recognize facts and laws? There are many practical situations when we fail to do so. They include intoxication with alcohol or drugs, tiredness and some mental diseases. We believe that our dreams do not represent reality. Contradictory testimonies in courts dramatically illustrate the unreliability of our observations. Large religious communities disagree about facts from the lives of their founding figures and more recent miracles. The famous experiment in which people overlooked a gorilla in the middle of a basketball game (see "Invisible Gorilla Test" in [49]) is yet another case of disconnection between reality and our "knowledge." Going back to falling apples, their downward motion is a "fact" only because we assume that we can trust our senses and information processing in our mind.

To apply induction and make a prediction we have to assume that the universe is stable and that we can correctly recognize the relevant facts and laws. These are the same assumptions that we make when we talk about "known facts." Logic cannot justify induction because induction does not work in some universes that we can imagine without violating the rules of logic.

One could wonder whether the asymmetry between the known and the unknown is a matter of degree. Is it true that predictions require a stronger form of the two assumptions than what we need to recognize past facts? To put it differently, can we be more certain of the past facts than of predictions, even if we accept that both are based on unverifiable assumptions? I believe that we typically derive our certainty from a web of various facts and laws which support our belief either in a past fact or in a prediction of a future event. On the practical side, we are typically aware of the past facts that are well confirmed in multiple ways. Our attention is often drawn to those future events whose outcome cannot be predicted well using known facts and theories. This distinction between the past and the future is not universal, though. I am much more confident that the Sun will rise tomorrow than that Shakespeare wrote the dramas attributed to him (see [50]). The distinction between the known and unknown is a matter of the strength of the connection of an event to other events and laws of nature.

7.1.2 *Induction is a law of nature*

Declaring the problem of induction to be ill-posed and leaving it at that would be hardly a satisfying answer to the puzzle. I will propose a new formulation of the problem and a solution to the new version of the challenge.

The starting point of my analysis is the observation that induction does *not* work. More precisely, sometimes it works, and sometimes it does not. There are numerous examples of predictions that failed. Some of the best-known examples are weather forecasts. People like to make predictions concerning functioning of the society, including wars, political campaigns and economy. So many of these predictions fail that often, they are not taken seriously. Induction failed many times when people tried to understand laws of nature. Many people believed that the Sun went around the Earth, that the Earth was flat, and that heavier objects fell faster than light objects. On the scientific side, the theories of phlogiston [51] and luminiferous aether [52] proved to be false, as far as we can tell.

The standard version of the problem of induction asks for an explanation of why induction works. This is a misleading question. A much better question is why induction works in so many cases that it is an effective tool in science and everyday life. One can also ask whether all possible universes can support sentient beings that can develop science. observations of our own universe on microscopic scales and very large scales suggest the negative answer. Many physical systems are not sufficiently complex to support sentient beings using induction. So induction is a law of nature in our universe. I do not know a good reason why induction would have to work in all possible universes.

I will base my analysis of induction on the concept of "resonance," to be introduced in Sec. 7.2. My proposed "solution" of the induction problem is no more than a reduction. Let us recall a well-known case of reduction to illustrate the concept. In antiquity, people believed that celestial objects moved along circles (with the annoying exception of planets). One explanation was that celestial objects were perfect, and a circle was the perfect shape for a trajectory. Newton explained the circular trajectories in terms of gravitation. Newton's idea was only a reduction or push of the philosophical problem to a lower level because the existence of the universal gravitation had to be assumed as an unexplained fact. This does not mean that reduction was useless—quite the opposite, the theory of gravitation was later used to interpret astronomical observations, predict the existence and positions of new planets and design and launch satellites. Science seems to be an activity always looking to reduce the current set of unexplained assumptions and phenomena to a more fundamental level. Newton's theory was followed by a deeper analysis by Einstein, in his relativity theory. Currently, physicists are trying to probe nature even deeper; one of their attempts is the string theory. I hope that my theory of resonance, despite being "only" a reduction, could explain some phenomena, make new predictions and inspire technology, for example, computers and other devices with "artificial intelligence" capabilities. But even if the theory of resonance does not achieve any of these goals, it will serve as an alternative look at the problem of induction. I think that the analysis of induction as a law of nature generates answers that are more interesting than those coming from attempts to create an inductive logic. The philosophical analysis of probability can be regarded as a specialized part of the analysis of induction.

7.1.3 *Anthropic principle*

The anthropic principle is presented in this way by [53],

> The anthropic principle [...] is the philosophical consideration that observations of the physical Universe must be compatible with the conscious and sapient life that observes it. Some proponents of the anthropic principle [...] believe it is unremarkable that the universe's fundamental constants happen to fall within the narrow range thought to be compatible with life.

I believe in the following instance of the anthropic principle: "We necessarily live in a universe where induction works."

Nelson Goodman introduced the idea of "grue" (see [54]). An object is grue if it is green until January 1, 2050, and it is blue at all later times. The philosophical problem with grue is that all available evidence supports the claim that grass is grue as much as the claim that grass is (permanently) green. So why should we think that grass is green? My answer is that the natural laws of our universe made us think that grass is green. So far, we are doing quite well applying induction as formed in our minds by the evolution.

Some scientists (see [47]) believe that there are many universes, including some with entirely different laws of nature from our own. There is no way that one can prove that grass is not grue because there exist universes, real or imagined, where grass is indeed grue.

The incredible effectiveness and reliability of induction is partly an illusion. We celebrate all cases where induction works well, and we seem to underestimate the extent of failures in philosophical discussions. Induction fails, for example, in weather prediction. The reason for this failure is not (only) that scientists have not developed reliable methods for weather prediction so far. The failure is due to a special mathematical character of some systems that was given a technical name of "chaos." In particular, weather is "chaotic." Chaos can be considered a law of nature—to save induction, we call a case of non-induction a law of nature. Once we move away from science, it is hard to find stable laws, even of the chaotic sort. And yet nobody complains that induction does not work in our world.

7.2 Resonance

The process of knowledge acquisition consists of two main parts. The second part was the traditional focus of epistemology, so I will give this part of the process the label "logical." The existence and preeminence of the first part of the process will be discussed next. I will refer to the first part of the epistemological process as "resonance" because I believe it has some common elements with the eponymous natural phenomenon. Roughly speaking, resonance is a crude first filter in the process of acquisition of knowledge. The logical processing is the much finer second filter.

7.2.1 *Information and knowledge*

I have used the words "information" and "knowledge" informally, and I will continue to do so in the rest of the book. I will give a special meaning to information and knowledge in this section so that these concepts help us understand the phenomenon of resonance.

Many diverse forces act upon every physical system. Examples include gravitation and electromagnetic waves. It is customary to give labels to many of these forces, even if, in principle, they are examples of or can be reduced to other physical forces. For example, we speak about friction, air resistance, and visible light. The totality of forces acting on a system can be called "information."

Knowledge is an ability of a system to process information in a way that is beneficial to it. The "beneficial" aspect of knowledge is quite subtle. When it comes to primitive animals, "beneficial" should be understood most directly. But the beneficial aspect of knowledge can apply to the whole species, or it can be interpreted as an improvement of DNA survival probability. It is even more subtle when it comes to humans, where knowledge can be used for purposes that are destructive or evil. Then knowledge benefits the goal set by a person.

The knowledge possessed by primitive organisms (animals and plants) is beneficial in the strict sense of the word because these organisms developed the simplest knowledge of the world via Darwinian evolution and selection. In this context, knowledge is the ability of the primitive organisms to react to the simplest chemical, light and gravitational stimuli in a way that increases their chances of survival and reproduction.

The functioning of the most primitive organisms is the simplest example of resonance. Simple organisms developed senses that can react only to some of the forces acting upon them. Resonance is the ability to filter only useful information from the enormous amount of the physical stimuli acting on a system.

7.2.2 *Resonance complexity*

I believe that a broad spectrum of phenomena can be reasonably classified as instances of the same general process of resonance. To put some order within this family of phenomena, I will propose a classification of resonance types based on the complexity of the stimulus and the action following it.

At the lower end, resonance detects a simple signal from the environment. Consider, for example, a primitive living organism that can react to light and move appropriately. An equally simple example of resonance is provided by (man-made) lights which are automatically turned on at night by a photosensitive switch. The stimulus can be simple even though the reaction can be very complex; for example, a single sensor measuring the depth of water may trigger a major response to the danger of flooding. Even when the signal is simple, the process that leads from the stimulus to the reaction may be very complicated. The route from a light stimulus to the motion of an organism is a complex chain of biochemical reactions no matter how primitive the organism is.

I am not sure how many different levels of resonance complexity one can recognize. I will indicate a few levels that seem to be distinct.

Animals react to danger. Danger can take different forms. Depending on the animal and circumstances, it includes cliffs, fire, and predators. Recognizing these dangers requires more than detecting a single type of stimulus.

At the next level of complexity of resonance we find reactions to social situations. Animals can recognize complex social cues and react appropriately. Social behavior requires an ability to detect signals with complex structures and process them in a sophisticated way. Humans can react to social cues, and we believe that our level of sophistication in this area is much higher than that attained by any other species.

The highest level of resonance complexity has been attained only by humans. In this case, resonance is the reaction to and processing of a very diverse and large body of information about the world. It involves sophisticated concepts such as marriage, democracy, war, God, and climate.

To do science, people have to recognize facts and construct theories, known in statistics as models. They have to select a small number of viable theories (models) because the number of theories is infinite, and one cannot analyze them all. The ability to select viable theories makes people much more intelligent than artificial intelligence at this time. People have a picture of reality in their minds and can generate models that are likely to be true within this picture. The best scientists can propose theories that go well beyond ordinary science but still have a good chance of success. Generating scientific theories (for example, statistical models) seems to be the highest form of resonance because of its complexity.

7.2.3 *Facts*

Our ability to recognize "facts" is the most elementary but fundamental example of human resonance.

Facts do not exist in isolation. A fact is a detected pattern. To say that a table is white is to say that an object is like other objects which we recognize as tables (that is, we recognize a symmetry in the universe) and it is also like other white objects (another symmetry). The representation of a fact as a pattern is especially significant in the social context. We do not know whether other people perceive tables as tables and white as white, but we can match our patterns of behavior (including speech) with those of other people, and thus we can agree on facts.

A fact is a distilled piece of information from an ocean of sensations, matched with memories in the process of resonance. Hence, the ability to determine and communicate facts is nontrivial and is subject to learning and failure.

It is quite often the case that failures help to illustrate success. Many machines are equipped with sensors and can perceive light and temperature and detect motion. All such equipment ages and fails at some point. People "fail" in similar ways. Some people are blind or deaf, limiting their perception, but they can still use other senses to compensate for these deficiencies. A more subtle and instructive case is people with autism. They fail to notice some social facts and cues—these are fundamental to the functioning of society.

Facts in our minds are not necessarily objective facts. Resonance is a part of information processing that takes input and generates output. The evolution made us, animals, good information processors in the sense that we can individually survive and multiply as species. This success does not prove that facts in our minds match facts in the objective universe. Quite the opposite, there are strong indications that our image of the universe may be incomplete or even distorted. Physicists consider quantum physics and relativity theory very solid scientific theories.

Nevertheless, our intuitive understanding of facts differs considerably from "facts" in these theories. On top of that, physics is ripe with more speculative theories that talk about higher dimensions, multiple universes, etc. Our perception of and belief in "facts" developed for the benefit of entities of our size and need not be universal. There might be sentient beings on other planets processing information in a different way. Just like religion evolved to help the survival of individuals and communities despite its distortion of facts, our fact perception might have evolved to help survival of large communities of cells (human bodies). None of this means that we should abandon our belief in facts. Our civilization rests on the foundation of facts, and nothing can change that. No matter how artificial our mental facts are, we must rely on them. The other option is uncompromising skepticism, with little theoretical or practical appeal.

The fundamental mistake of de Finetti and some positivists was that they thought that scientists could recognize facts and only facts. There exists a view of science according to which science collects facts and then arranges them into theories. De Finetti thought observing facts was sufficient for applying probability. He failed to recognize the fundamental role of the human ability to recognize relationships. In probability, symmetry and independence are fundamental relationships. Actually, these two relationships are fundamental to all of science. Other relationships are also used, but their position is less fundamental than those of symmetry and independence. De Finetti implicitly accepted our ability to recognize one type of relationship—that some events are mutually exclusive. According to de Finetti, this is all you need to match Kolmogorov's axioms with reality. Von Mises' probability theory was based on our ability to recognize collectives. Von Mises missed some other relationships by restricting his concept of probability only to collectives.

7.2.4 *Learning resonance*

The most primitive animals cannot learn much at the level of an individual. Their learning process consists of mutation and selection at the species level. Due to this process, consecutive generations respond better and better to the environment. The higher we go on the ladder of life, the more developed is the process of learning at the individual level. It seems that the ability to learn at the individual level is one of the main reasons for the crushing success of the human population relative to other animal populations. A well-known proverb says, "Give a man a fish, and you feed him for a day; teach a man to fish, and you feed him for a lifetime." The evolution made a big step forward when it equipped humans with an excellent ability to learn at the individual level, hugely exceeding the learning abilities of even the most advanced apes. Humans developed a whole system of learning for individuals based on schools and universities but much, perhaps most, of the individual learning is informal.

A part of the learning process, both in animals and in humans, is learning resonance. This consists of moving learned skills from the conscious level to the subconscious level. Polanyi points out in [55, p. 49] that the human ability to ride a bicycle is mysterious. Although this ability can be described in abstract scientific terms and, to a certain degree, it can be explained to people learning how to ride a bicycle, the learning process and the ability are mostly unavailable to scientific analysis at this point. The cyclist's ability to detect changes in the position and velocity of the bicycle, a prelude to taking corrective turns of the handlebar, is an example of learned resonance.

The part of the process of learning of resonance most relevant to my analysis of probability belongs to the formal education system. Students of all ages listen to theoretical explanations and practice related skills by doing numerous examples. Both [56] and [55] point out that the learning process is based on practicing a skill by doing many examples. Understanding the theoretical principles underlying all the examples comes later. The structure of most traditional and current textbooks is the opposite—the presentation of a topic starts with the theory and is followed by examples.

A successful learning process turns even complex theoretical concepts and advanced skills into instances of resonance. Students who mastered the material do not have to "think" about how to apply the theory which they have learned. The practice establishes the corresponding skills in the subconsciousness. From now on, students apply the learned skills and ideas according to the principles of resonance—instantaneously and subconsciously. The ability to read is a good basic example. Reading a text is not processing individual letters but processing whole words or even sentences.

The learning process builds new resonance abilities based on already established abilities. Let us have a look at an example of such a hierarchy in learning mathematics. Students first learn fractions. This skill is used to learn differentiation and integration. These skills, in turn, are the basis for learning how to solve differential equations.

People are trained to recognize very complex facts because it is virtually impossible to pass directly from simple facts to successful decision strategies either in science or everyday life. Scientists and other professionals must recognize, for example, sulfur, lasers, supernova explosions, and cancer.

Kuhn's "scientific revolutions" (see [56]) are, among other things, changes in resonance abilities. Scientists learn resonance according to a theory and resist the change for practical reasons—they find it hard or impossible to apply resonance according to a new theory. In more elementary language, they lack intuition needed for applications of the new theory. This may be one of the reasons for the resistance to change and the need for revolutions to replace old theories with better ones.

Kuhn recognizes the vital role of "normal" science, but he seems to underestimate its training role. Learning science is far from learning facts from textbooks. Scientists have to develop resonance abilities at the highest level. The only practical way of learning resonance at this level is to conduct "normal" research.

The scientific laws of probability (L1)-(L6) are the starting point for the teaching of probability. At this level, most examples in probability textbooks are combinatorial. They often deal with highly symmetric objects and situations, such as coins, dice, playing cards, etc. The remaining probabilistic material is taught in the hierarchical style mentioned before. For example, students learn at some point about the normal distribution. Then they learn about Gaussian processes. This can be followed by the theory of stochastic partial differential equations (SPDEs) with Gaussian solutions. Each of these topics relies on the previous one.

7.2.5 *Resonance level reduction*

Resonance level can be reduced in practice. In a statistical context, if someone cannot recognize that observations come from a normal distribution at the intuitive level, there are methods to check if the distribution is normal. This process requires some data and some theory, so it is applied only in those situations where it is really needed. However, the resonance level cannot be reduced in some other cases. A good practical example of irreducible resonance are sequences called "i.i.d." (independent identically distributed) by some statisticians and "exchangeable" by others. One has to recognize a symmetry in this case—further reduction of resonance level is not possible. Statisticians have methods that can be used to determine whether a given sequence is i.i.d. or exchangeable but these methods are based on the ability to recognize some other symmetries (i.e., facts and models). This illustrates the possibility of moving the target of resonance within the same level of complexity to achieve a practical goal.

A standard method to reduce the resonance level, both in everyday life and scientific practice, is to use a measuring instrument such as a thermometer or scale. These instruments act as resonance "antennas" in that they filter the relevant information from the environment. Then, they convert it to the easily human-recognizable form using a lower-level or more reliable resonance form. A measuring instrument must be based on a theory—a set of assumptions about how the world works.

7.2.6 *Properties of resonance*

Examples of resonance in previous sections should have helped the reader understand the concept. I will now list the essential attributes of resonance to clarify its meaning.

(i) Resonance is a crude filter of information.
(ii) Resonance is the first step in the knowledge creation process. It comes before logical analysis (if there is any).
(iii) Low-level resonance is instantaneous. It is the fastest of all information processing steps. High-level resonance (for example, some resonance involved in scientific research) need not be fast—see Sec. 7.3.
(iv) Resonance is atomic. It is not a process consisting of steps.
(v) Resonance resembles a "finite-dimensional projection" in mathematical terminology. No matter how small a system is, all forces acting on the system can be reasonably represented as an infinite dimensional family of stimuli. Resonance cuts dimensionality to one or a few dimensions. Most of the potentially available information is discarded.
(vi) Resonance, especially at the higher levels of complexity, detects information that is not normally considered to be elementary in science, in terms of the fundamental forces of nature. Quite the opposite, at the highest level of resonance complexity, humans instantaneously detect anger, happiness and aggression. In the statistical language, this resembles the detection of "principal components" which are combinations of simpler quantities (stimuli).
(vii) Resonance is not conscious. People are aware of the resonance results and can be aware of the resonance. For example, they may notice their detection of someone's anger and

they are aware of their reaction to someone's anger. But people have no insight into the nature of their resonance and cannot analyze this part of their information processing. This does not mean that people cannot correct the results of resonance—logical and scientific analysis is the second, more accurate step of information processing. People are also capable of manipulating their resonance. For example, some people make a conscious effort to eliminate their racial prejudices.

(viii) The results of resonance, especially at the high level of complexity, come with a feeling of their (high or low) reliability. This contrasts with the lowest level resonance, which is either acted upon or not. The feeling of the degree of reliability is what many people would call "subjective probability." I object to this misnomer because the feeling of reliability is informal but not arbitrary. The informal reliability assessment is crucial to intelligent information processing. The results of resonance that are considered somewhat unreliable can be subject to logical processing. They can also inspire efforts to collect more relevant information and thus generate a more reliable assessment of reality.

7.2.7 *Physical basis of resonance*

I do not know the physical basis of epistemological resonance. It is possible that scientists working on human brain have already determined the physical basis of resonance or will determine it soon.

It will not be much of a surprise for anyone that I have chosen the label "resonance" for the epistemological process because I believe that it might be related to a physical phenomenon called "resonance."

Resonance occurs when a system reacts to a periodic force. For resonance to occur, the period of the force must belong to a restricted family of numbers characterizing the system. Radio receivers are designed to pick up the signal of only one frequency at a time. All other frequencies in the electromagnetic wave spectrum are ignored. Undesirable resonance occurred in some faulty engineering designs, for example, some bridges and airplanes. Some people can break a wine glass by singing a note that perfectly matches the fundamental frequency of glass vibrations.

On the mathematical side, physical resonance could be represented as filtering of all eigenfunctions except the first one (or perhaps the top few eigenfunctions). In other words, resonance is a form of mathematical projection. One could speculate that the epistemological resonance is, at least sometimes, based on a similar projection on an eigenfunction subspace in some high-dimensional or complicated space. I apologize to readers without scientific background for the technical jargon used in this paragraph. I can assure you that this is no more than speculation—I have no solid evidence linking epistemological resonance to these mathematical concepts.

Our low-level resonance abilities are limited to events which are manifestations of physics laws governing objects at our magnitude. We can access other levels of physical universe, for example, quantum and relativity phenomena, via experiments involving resonance (observations) at our level. It is an interesting question whether other sentient beings could observe facts represented by physics theories for very small or very large objects in a direct way using some other resonance capabilities. I guess that this could be the case if not in ours, then perhaps in other physical universes.

There has been much talk about and some progress on quantum computing in recent years. The advantage of quantum computing over digital computing is that some calculations are done in parallel, in an infinite dimensional space, in a sense. Resonance has a similar advantage over logical processing of data. In the resonance process, a single mode can be filtered in a single step from data that are infinite dimensional. In other words, we can instantly and intuitively judge which data (aspects of reality) are irrelevant.

A less sophisticated way to speed up computations is parallel computing. I do not think that parallel computing is a good analogy for resonance, in contrast to quantum computing. Parallel computing shortens the computing time by assigning the same task type to multiple processors.

Recent AI (artificial intelligence) advances based on "large language models" suggest that computers can acquire low-level and medium-level resonance skills. There is a debate on the limits of AI. From the perspective of my resonance theory, this can be phrased as a question of what level of resonance neural-network programs trained on the contents of the Web will be able to achieve.

7.2.8 *Resonance and subjectivity*

Resonance is suspect, but it is not subjective or irrational.

My main motivation for developing an epistemology, and especially the theory of resonance, is my desire to understand the subjective component of the concept of probability. I believe that the main reason why people call probability subjective is a relatively simple case of confusion. However, one cannot dismiss summarily all claims about the subjectivity of probability made by people who devoted much thought to the problem. I came up with the concept of resonance because this is the only way, in my opinion, in which the claims about the subjectivity of probability can be made at least a little bit rational.

Calling resonance subjective would do more harm than good. I firmly believe that the word "subjective" has connotations of freedom. In my opinion, neither Mona Lisa was beautiful, nor the painting of her is beautiful. I do not like honey, beans and whiskey. I like red wine, but I do not like white wine. I like jazz and opera, but I do not like country and Western music. All of the above are my subjective opinions. I identify the subjectivity of these opinions with freedom because I feel that I can hold these beliefs no matter what others think. It is unlikely that I will suffer any consequences for expressing these beliefs. I can change these beliefs at will and without justification (in fact, I changed some of them at some point in my life).

Resonance does not resemble or involve freedom in any conceivable sense. Resonance refers to the fact that a part of our mind (or that of more primitive animals or machines) acts as a black box. It accepts some input and generates some output, but we cannot examine the inner workings of the black box in real-time. The black box may be very accurate or not, but it is not arbitrary. We mostly care about those cases of resonance that seem very accurate. From the practical point of view, people can recognize basic facts with great accuracy despite all types and cases of failure. Modern computers act like black boxes for most people because most people know very little about the details of computers' hardware and software. However, nobody would apply the word "subjective" to the information processing done by computers.

I think the word "suspect" characterizes the resonance phenomenon better than the word "subjective." Since we do not have easy access (or we have none) to the process of resonance, it is automatically suspect, on the purely intellectual side. We can only develop some degree of trust in our resonance ability by observing many successful resonance applications. Of course, this is somewhat circular because "observing success" requires recognizing facts, which requires resonance. But one can say that resonance fits well into a certain logically consistent theory of the universe.

7.2.9 *Raw resonance*

Resonance can be wrong. This is not necessarily bad news—let me start with a few positive examples. Color television and computer displays are based on the successful simulation of colors; this is a polite way of saying that our senses are fooled. Color TV and computer monitors display only mixtures of three colors: red, green and blue. We have an illusion of seeing all possible pure colors in the spectrum and all their possible mixtures.

Some people can easily tell the difference between the colors on a computer screen and the colors of a real object, but this ability varies from person to person. Not only TV and computers are based on the same mental inability to distinguish precisely between various mixtures of primary colors. This phenomenon is behind the successes of color photography, color print and even the venerable art of painting. Note that painters have only a finite number of dyes to choose from. If we pursue this direction of thought, we realize that the success of the classical painting is based, in part, on our ability to "see" three-dimensional objects in a two-dimensional plane. Likewise, the film is based on our inability to separate individual frames into discrete sequence—we see a continuous motion on the screen.

Next, we go to the dark side. According to [57], "Approximate statistics on the number of trials for witchcraft and executions in various regions of Europe in the period 1450–1750" give a lower estimate for the number of executions of witches at 35,000. About 75 to 80 percent of those were women. Witch-hunt was a complex social phenomenon with multiple roots, including economic, political and religious causes. But, at a certain level, every individual case involved resonance—the ability to recognize a witch in a more or less direct way from the available empirical evidence (including gossip, etc.). I do not believe that all people involved in witch-hunt (on the side opposite to the "witches") were evil or had hidden motives. I think that many of them genuinely believed in their ability to recognize witchcraft and witches.

There were innumerable cases of false beliefs of various caliber. I have chosen witch-hunt as my illustration because of its drastic character and clear, from the modern perspective, lack of understanding of the laws of nature. I am sure that the scourges of racism, antisemitism, misogyny, etc., involve some dose of mistaken resonance.

The evolution created a human propensity to recognize "facts" via resonance in many areas of life. In some of the areas, the "facts" are seldom confirmed by the scientific process of empirical verification. Religion comes first to my mind as an example of this tendency, but it is not the only example. Political ideologies enjoy significant loyalty. Many people tend to follow and believe in strong leaders.

Since resonance is often misleading, how do we know it is not equally misleading in scientific research? Actually, resonance is misleading in science. Our normal instinctive picture of the physical world agrees with Newtonian physics. This is a false belief according to quantum mechanics and theory of relativity. Newton's physics is the most effective way of interpreting the world for human-sizedhuman organisms. Resonance does not necessarily lead to the truth about the world. Most of thc time (we hope), it leads to the most beneficial interpretation of the world from the point of evolution.

7.2.10 *Is resonance a new concept?*

Is resonance just a new name for an already well-established concept? There are several candidates for a precursor of resonance: subconsciousness, filtering, pattern recognition and instinct.

One could say that resonance is subconscious information processing, and logical reasoning is conscious information processing. This identification is a useful rough idea but not totally accurate. I do not know the full extent of subconscious information processing (and it appears that nobody knows at this time), but subconsciousness is certainly not limited to resonance, that is, a specific and rather narrow in scope process of filtering of relevant information. Similarly, conscious information processing does not have to be limited to what we call logic.

The epistemological resonance (that is, the resonance in my sense of the word) is closer to the physical resonance than to signal filtering—this justifies my choice of terminology. Filtering signal from noise presupposes the existence of signal and noise. At the highest level, resonance includes the human ability to discern patterns in the available information. This applies equally to witty remarks at a party and new proposals for historical theories. Although the posterior analysis of such resonance cases could represent them as signal filtering, I doubt this does justice to the idea of resonance. We can classify components of an information stream as signal and noise, either if we understand the source of the information and noise, or as a result of the a posteriori analysis.

I do not think that we can identify resonance with pattern recognition. It is conceivable that all resonance can be represented in some abstract way as pattern recognition, but I doubt it is useful to think about resonance this way. Some resonance forms are much more complex than what we normally call pattern recognition.

Instinct is an instance of resonance, but the concept of resonance is broader than the concept of instinct. We typically talk about instinct only when it is innate; that is, it is not acquired via learning. I do not limit resonance in this way because the origin of resonance abilities does not necessarily affect resonance effectiveness. It is best to stretch the concept of resonance well beyond the domain of behavior typically referred to as instinctive. Thus, I include the simplest reflexes of the most primitive organisms and the most impressive examples of human intelligence in the concept of resonance. The uniting element of these examples is the lack of logical or complex information preprocessing and the highly selective and successful stimuli filtering.

Two types of thinking have been described by Kahneman in [58]: System 1 (fast, automatic, frequent, emotional, stereotypic, unconscious) and System 2 (low, effortful, infrequent, logical, calculating, conscious). While resonance shares characteristics with System 1, its essence is philosophical rather than practical. Resonance connects people to reality. In some cases, resonance is closer to System 2 than System 1 (see the next section).

7.3 Intelligence

I see a close relationship between intelligence and resonance. Intelligence is primarily an ability to apply resonance at the highest level. This ability comes at different sublevels and occurs in a variety of situations. We apply the term "intelligence" to those instances of resonance that stand out in quality and complexity. Animals apply resonance to recognize objects and react accordingly. So do humans. People apply resonance in much more demanding contexts, for example, to quickly recognize social messages conveyed by facial expressions.

Examples of intelligent behavior include the ability to recognize investment opportunities. Some people react quickly to the slightest cues of social nature. Yet other people may be able to notice natural phenomena that are not so apparent to others.

The kind of intelligence that is most relevant to this book is the ability to construct new scientific theories. Although the story about an apple falling on Newton's head is not entirely accurate (apparently he had a great idea sitting under a tree; there is no evidence that an apple fell on his head), it illustrates well intelligence as a form of resonance. All people could see the Moon moving in the sky and apples falling from trees. However, only Newton realized that the two phenomena might have something in common. Einstein likewise did not base his relativity theory on his own new experiments or observations. Many physicists knew the same scientific theories and facts that inspired Einstein's theory. Newton and Einstein had antennas in their heads, able to resonate at the frequencies of fundamental laws of nature. In this sense, they were exceptionally intelligent.

The simplest type of induction consists of drawing conclusions or conjectures from repeated observations of the same event. An example of intelligence is to draw a conclusion or a conjecture from observing a single event. This could be a coincidence—the simultaneous occurrence of two unrelated events, especially if they are both highly improbable. Simple mechanical and biological systems cannot acquire knowledge in this way because an intelligent agent has to have a model of the world in his mind, according to which both events have small probabilities. Moreover, he has to single out only two events from billions of irrelevant facts. Finally, he must be able to come up with an alternative explanation, linking the two events in a single theory, which gives reasonably high probability to the observed "coincidence."

People are inclined to see coincidences where there are none. This is typical of superstition—a good example is astrology. Science has to weed out false theories based on (real) coincidences. It also needs intelligent (and perhaps lucky) scientists who can notice coincidences and conjecture theories whose reasonably large proportion will turn out to be (approximately) true.

Since finding new significant scientific theories is challenging and cannot be easily taught or programmed, society puts a premium on new, interesting ideas. The discoverers of new significant theories are rewarded by prestige or money. This creates a problem. Some people create highly speculative theories, hoping they will prove correct and earn the inventor recognition. This trend is present in modern physics, parts of which have become very speculative. One can see the same trend in an area as distant from science as modern art where artists try to outdo themselves in inventing more and more strange forms of modern art.

Most science is "normal" in the sense of [56] and not revolutionary like that of Newton and Einstein. Constructing scientific models is sometimes routine, and one can imagine that properly programmed contemporary computers could play this scientific role. But routine science, no matter how much it is needed and appreciated, does not draw the greatest admiration. Nobel prizes go to people who applied intelligence (resonance) in ways that are well beyond the capabilities of most other scientists.

Having poured praise on geniuses I have to point out that normal science needs normal intelligence and normal resonance. Mainstream scientific theories are not invented due to logical reasoning, although once they are invented, they are often justified (in a preliminary way) using logical arguments bringing many ideas together.

I have proposed in Sec. 7.2.6 that resonance is very fast, almost instantaneous. Intelligence is an exception to this rule on some occasions. Simple resonance is unquestionably fast—we react quickly to dangers such as fire, falling objects, etc. But the "eureka moment" might build on a long, most likely subconscious, process leading to the final discovery. Some scientists reported a period of incubation for their ideas. At this point, scientists are not able to observe mental activity in the brain with sufficient accuracy to determine the length and structure of processes leading to intelligent behavior.

Chapter 8

Science

I will limit my analysis of science to several topics that are not necessarily the most significant but are most relevant to the subject of this book. I have already related science to resonance in Sec. 7.3. I will refer to this relationship occasionally throughout this chapter and the remaining part of the book. I will analyze science as a communication system because I consider my scientific laws of probability (L1)-(L6) to be the de facto basis for the communication system for the probability and statistics ideas. I will also argue that science is a defense against intellectual fraud because this topic is closely related to the claims of subjectivity of probability.

8.1 Science as a communication system

A unique feature of humans among all species is our ability to communicate using language. Individuals of many other species, from insects to mammals, can exchange some information between each other, but none of these cases comes even close to the effectiveness of human oral and written communication. The language gives us multiple sets of eyes and ears. Facts observed by other people are accessible to us via speech, books, radio, etc.

The wealth of available facts is a blessing and a problem. We often complain that we are overwhelmed with information. A simple solution to this problem emerged in human culture a long time ago—data compression. Families of similar facts are arranged into patterns, and only patterns are reported to other people. Pattern recognition is not only needed for data compression. It is also the basis of successful predictions. People generally assume that patterns observed in the past will continue in the future, so knowing patterns gives us an advantage in life. An important example of "patterns" are laws of science. Some people are not as good at pattern recognition as others, so communication gives them not only access to multiple sets of eyes and ears but also access to multiple minds. Communication enables the exchange of information not only between concurrently living individuals but between generations as well. The result is accumulation, one could say explosion, of knowledge, including scientific knowledge.

The process described above is not perfect. Our senses are imperfect, our memory is imperfect, and our ability to recognize patterns is imperfect. On top of that, communication adds its own errors. Some of them are random, but some of them are typically human. What we say may be colored by our political or religious beliefs, for example. Some people pursue their goals by spreading misinformation, that is, they lie. Experience taught people to be somewhat skeptical about information acquired from other people. Information is categorized, and different batches of information are considered reliable to a different degree. Science may be defined as the most respected and most reliable knowledge that people offer to other people. The distinguishing feature of science is its method. Science achieved its high status in various ways, for example, scientific claims are often repeatedly verified, the ethical standards imposed in science are much higher than in politics, assault on established theories is approved, facts rather than feelings are stressed, the simplest theory is chosen among all that explain known facts, etc. Religion seems to lie at the other extreme of major ideologies. The utterly counterintuitive claims of the quantum theory and relativity theory are widely accepted by populations as diverse as democratic societies and communist societies, Catholics and Muslims. On the other hand, humanity seems to have reconciled itself to the coexistence of various religions without any hope for the ultimate coordination of their beliefs. In other words, religious information conveyed from one person to another may be met with total skepticism, especially if the two people are followers of different religions.

To maintain its elevated status, science has to present facts and patterns in the most reliable way. The most general patterns are called "laws" in natural sciences. The history of science shows that we cannot fully trust any laws, for example, the highly successful gravitation theory discovered (or invented) by Newton was later fundamentally revised by Einstein. The laws of science are the most reliable information on facts and patterns available at this time, but they are not necessarily absolute truths.

The success of science (and human communication in general) depends very much on the universality of certain basic perceptions. In other words, almost all people agree on certain basic facts, such as numbers and colors. When I look at five red apples, I am quite sure that any other person would also see five red apples, not seven green pears. Of course, we do make counting mistakes from time to time. The further we are from numbers, the harder it is to agree on directly perceived facts. If two people cannot agree on an answer to a question such as "Is the water in the lake cold or warm?", they can use a scientific approach to this question by translating the problem into the language of numbers. In this particular case, one can measure the water temperature using a thermometer. Numbers displayed by thermometers and other measuring devices are a highly reliable way to relay information between people. However, one must note that no scientific equipment or method can substitute for the prescientific agreement between different people on some basic facts. We must assume that our sensations generated by resonance are analogous to those generated by the resonance process in other people's minds. For example, suppose that a distance is measured and it is determined that it is

8 meters. A person may want to communicate this information to another person in writing. This depends on the ability of the other person to recognize the written symbol "8" as the number "eight." The problem cannot be resolved by measuring and describing the shape of the symbol because a report on the findings of such a procedure would contain words and symbols that another person might not recognize. The example seems to be academic, but it is less so if we think about pattern recognition by computers.

One of the main reasons for the success of natural sciences is that most of their findings are based on directly and reliably recognizable facts, such as numbers, or they can be translated into such language. Measuring the spin of an electron is far beyond the ability of an ordinary person (and even most scientists), but the procedure can be reduced to highly reliable instructions, and the results can be represented as numbers. The further we go away from natural sciences, the harder it is for people to agree, partly because they cannot agree even on the basic facts and perceptions. A statement that "Harsh laws lead to the alienation of the people" contains the words "harsh" and "alienation," whose meaning is not universally agreed upon. A very precise legal-style definition may be proposed for these words, but such a definition need not be universally accepted.

I used Feyerabend's book [31] in Sec. 4.1.3 to illustrate the pitfalls of skepticism. The book deserves another honorable mention for the incredible distortion of science. In Chap. 17 and many other parts of his book, Feyerabend claims that some scientific theories are "incommensurable" with others. This means that the claims of pairs or families of theories are neither in agreement nor in contradiction. Feyerabend is wrong—the essence of science is that it is commensurable. People must have always been aware of the frequent incompatibility of their observations, arguments, opinions, etc., with those of others. Some philosophers represented the discovery of the scientific methodology as a new relationship between people and nature. At least as significant as that development was that science was a new relationship between people. Because of science, large groups developed trust in one another concerning various matters of fact and theory. All Feyerabend's examples of incommensurability were extracted from the fringes of science. They came from antiquity, from the infancy of some branches of science and sciences under development, such as medicine. Incommensurability is a figment of imagination—every person with at least an average education knows that copper is a good conductor of electricity and glass is not.

8.2 Some attributes of science

Interpersonal character

Science is a set of facts that can be described and theories that can be implemented by everyone. These collections span the spectrum from sensory observations to advanced scientific theories that have been determined to be reliably intersubjective. This means that information passed on from one person to another can be reliably verified by the receiving side. Of course, a single person does not have time, energy, knowledge, etc. to verify all scientific claims. Many advanced scientific claims can be verified only by a handful of highly trained experts. But every scientific claim can be verified by everybody in the following sense—any person or organization with sufficient resources (money, time, energy, materials, buildings) can hire people who have the relevant expertise and can verify the claim. Good examples of collections of advanced claims are modern physics and its technical implementations such as space rockets, smartphones and nuclear plants.

The elements of science that are interpersonal and reliable are not necessarily as basic as counting or recognizing colors. Highly trained experts can easily recognize more complex objects or situations such as zebras and supernova explosions. In other words, resonance comes at different levels.

The role and limitations of resonance

Scientific theories are often presented as models with parameters. The model is obtained through the process of resonance. It can be analyzed by the usual scientific process of falsification/verification based on logic. The space of models is discrete, in a sense. If a model is rejected, there is no presumption that a better model for the same phenomenon should be "close" to the original one, although it could. The parameters turn each model into a vector space or manifold (a continuous structure). The process of resonance is not sufficiently effective to provide the values of the parameters. But resonance gives bounds on the values of parameters. Consider Newton's Law of Universal Gravitation. It states that the force of attraction between two bodies with masses m_1 and m_2 at the distance r is equal to Gm_1m_2/r^2. Here G is a constant that is a parameter. Its value can be obtained from experiments but not from resonance. The number 2 in r^2 may be interpreted as a parameter to be measured in experiments or as a resonance-derived constant because 2 is the only power of r which makes some systems based on gravitation stable.

Convergence to the truth?

I wish I could say that science gets closer and closer to the objective truth about reality but my skepticism does not permit me to go that far. It is interesting to notice that neither Kuhn nor Popper, the two best-known philosophers of science in the twentieth century, claimed that science brought us closer to the truth. Science is an attempt to understand reality using a particular set of intellectual tools. The tools are designed to select these claims that are practical and also to get rid of manipulation by unscrupulous individuals and ideologies. The practical success of science does not prove that it is a faithful representation of objective reality.

Science control

An important characteristic of scientific research is that it can be directed to and focused on the most significant topics, gaps, anomalies, etc. This gives science a huge advantage over those fields and practices where information is collected randomly or arbitrarily.

Asbestos, lead and cigarettes are examples of substances and products that have been scientifically proven harmful. Since powerful industries made profits from selling these products, misinformation and political influence were (and in some cases still are) used to keep the manufacturing of these products legal. Science used one of its methods to establish the truth about these substances' harmful effects—overwhelming amounts of data were collected and analyzed by independent researchers.

Religion is an example of ideology where directed or controlled empirical research is not practiced. These activities seem incompatible with the nature of religion.

Economics is a good intermediate example. Economists collect enormous amounts of data, but they have limited ability to perform controlled experiments, especially on the "macro" scale.

Beyond simple induction

The effectiveness of science is due to reliance on scientific theories. They enable scientists to make predictions in situations that do not repeat. In other words, theories let scientists make reliable predictions that are not directly based on repetitive observations or experiments.

Technology is a good illustration of this feature of science. The first airplane was a scientific prediction, in a sense. The Wright brothers "predicted" that the machine they constructed would fly. Needless to say, that prediction was not based on a long sequence of observations of similar flying machines. The design was based on repeated observations of some phenomena; for example, the plane had wings like birds. Also, the Wright brothers did experiments with various propeller designs. Nevertheless, the airplane itself was not identical to and not even very similar to any other object.

Science as a web

Science is a web of highly interrelated facts, theories and fields. For example, the evolution theory is a field of biology. Still, it has to agree with biochemistry (via research on DNA and related molecules), geology (dinosaur bones) and astronomy (evolution of the solar system). In contrast, religion quite often offers isolated claims, such as the Creationist claim that the universe was created in its present shape in the last 10,000 years. Changing a significant part of the evolution theory would affect many related theories. Creationists can easily adjust their claim and move the purported date of creation without the danger of creating a logical inconsistency.

The necessity to coordinate diverse scientific theories creates an opportunity for multiple attempts at falsification of assertions. This also applies to predictions made using probabilistic methods.

8.2.1 *Science as an antidote to manipulation*

Science is partly a result of developments in information processing in societies. First, people learned how to communicate, pass information on to others, receive information and use information to their benefit. Then, they noticed that they could lie and manipulate other people. And then they noticed that other people lie and manipulate. Science is an answer to this problem—a system that ensures that the effects of other people's lies and manipulation are minimized. This system has many elements.

(i) Scientific theories are supposed to be expressed in terms that either are or can be reduced to those facts that are most widely recognized as true. These are variably referred to as qualia, basic facts, sense perceptions, etc.

(ii) Most successful and solid scientific theories are based on a multitude of experimental evidence. Science has methods of improving the reliability of observations and measurements.
(iii) Science is expressed in terms that give clear operational instructions to people who want to verify factual and theoretical claims. Some theories are too advanced to be verified, and some experiments are too costly (in terms of money, energy and labor) to be performed by an ordinary person, but they can be repeated at least by experts or teams of experts.
(iv) Publishing false scientific claims results in a level of opprobrium far exceeding the level of criticism of dubious claims made by representatives of other ideologies (political, religious, etc.).
(v) Despite the presence of all the usual evils of social interactions, such as conformism and suppression of criticism, in the scientific community, there is a great degree of support for new and even revolutionary ideas. Einstein and Darwin have been role models for generations of scientists. Searching for errors within established theories is a part of the scientific ethos.
(vi) Scientific claims are repeatedly tested. The standing of a theory depends on the severity and the number of tests that were performed or that can be performed. Hence, there are substantial differences between the reliability of physics, chemistry, biology, medicine, economy and humanities.
(vii) Science is a network with fractal growth, that is, it grows in unexpected directions, it is sometimes circular in the sense that it looks at the same facts and theories from different angles, magnifies some parts and fills some gaps. For this reason, trying to negate a scientific fact has the same effect as removing a card from a house of cards—the whole structure may collapse. It was suggested in [7] that the Moon is made of green cheese with strictly positive (personal) probability. If the cheesy composition of the Moon is ever confirmed, most of the contemporary science will collapse. At the other extreme, many key theological claims

can be modified individually without much harm to the rest of a given theology.

(viii) Prediction is stressed in science because it is hard, and so it is a severe test of a given theory. Explanations of existing facts are easier to come by and harder to eliminate by logical analysis.

8.2.2 *Dependent information sources*

Standard Bayesian information processing in our minds increases our belief in a claim if and when some new piece of supporting evidence arrives. The evidence may have the form of hearsay. It would not be wise to reject all indirect evidence. The problem with this mode of opinion-forming is that we are not very good at recognizing situations when multiple information channels are highly correlated. We are all bombarded by propaganda of a political, religious and social nature. Many instances of the same message are often highly correlated. Despite the high correlation of information sources, people feel that a claim is corroborated if they hear multiple expressions of the same opinion.

Science is not immune from this danger. But science is the only ideology which placed multiple independent verifications of every claim on a high pedestal.

If people were totally gullible and succumbed to all propaganda then the society would likely collapse. So a compensating behavior developed, and many people became immune to any amount of propaganda in some specific fields or coming from specific sources. There are several unfortunate results of these developments in the social communication system. The first one is the war between propaganda sources, which encourages information providers to use more and more extreme methods of communication and indoctrination. Another negative effect is a polarization within minds of individuals (but also groups of people). The whole universe of information is often divided into reliable sources, no matter how dependent and deceitful they might be, and unreliable sources and arguments, no matter how reasonable they might be. A common result is an almost complete ossification of opinions among some people and groups, especially on emotionally charged subjects, such as religion, politics and social issues.

Returning to science, one may say that science attempts to counteract these counterproductive practices and habits. Ideally, scientists should listen to all reasonable arguments, set aside emotions and loyalties, and test all claims multiple times, independently, and in diverse ways.

8.3 Technology

Polanyi presents in Sec. 6.8 of his book [55] a profound discussion of technology as an activity fundamentally distinct from science. I will make only a few remarks on technology. Animals apply simple induction just like people. Primitive animals "learn" beneficial behavior via the selection process (a part of evolution). More advanced animals learn at the level of individuals; that is, individuals can notice patterns and learn beneficial behavior as a reaction to repeated external conditions. But very few animals can make tools. This ability is limited to a handful of species, mostly apes and some birds. Moreover, animal tools are rudimentary. The unbridgeable gap between humans and animals is most apparent in technology. One does not have to point to the space shuttle to prove our superiority. Animals have not developed anything as simple as a cart. The highly complex structures built by ants, termites and some birds result from pure instinct.

Technology is based on scientific theories, not generalizations from repeated observations. Inventors build new equipment in their minds using novel applications and combinations of the known laws of nature. Technology applications, a form of positive verification, provide a much more convincing proof of the validity of a scientific theory than any laboratory experiment.

8.4 Decision making

The unique characteristic of statistics among all natural sciences is that decision theory is embedded in it in a seemingly inextricable way. I will outline my philosophy of decision making in relation to my philosophy of science.

Decision making is not considered a part of science in deterministic situations. For example, it is up to a chemist to find the melting temperature of gold (1064°C), but it is left to potential users of chemistry to implement this piece of scientific knowledge. If anybody needs to work with melted gold, he has to heat it to 1064°C. The decision to heat or not to heat a piece of gold is not considered a part of chemistry. The laws of deterministic sciences can be presented as instructions or logical implications: if you heat gold to 1064°C then it will melt. If you want to achieve a goal, all you have to do is consult a book, find a law which explains how to achieve that goal and implement the recipe. This simple procedure fails when a decision problem involves probability because the goal (the maximum possible gain, for example) often cannot be achieved with certainty. In decision theory, assuming that the decision-maker wants to maximize their gain is standard. If no decision maximizes the gain with certainty, the decision-maker has to choose among available decisions using some criterion not based on the infallible attainability of the goal. The choice is not obvious in many practical situations. Historically, statistics had a strong decision-theoretic component—scientists felt that it would be unfair to leave this matter in the hands of lay people who might not be sufficiently knowledgeable about the decision making.

The nature of the decision making problem is not scientific. Science can predict the results of different decisions, sometimes with certainty and sometimes with some probability, but it is not the business of science to tell people what decisions they should make.

The identification of decision making and probability assignments by the subjective theory of probability is misleading. The identification is only a mathematical trick. The subjectivist claim that your decision preferences uniquely determine your probabilities (and vice versa) refers to nothing more than a purely abstract way of encoding your preferences using mathematical probabilities. This part of the subjective theory shows only a mathematical possibility of passing from probabilities to decisions and vice versa, using a well-defined mathematical algorithm. If probability is objective, it is not evident at all that decision preferences and probabilities should be identified (see Chap. 10).

Chapter 9

Science of Probability

The science of probability must provide a recipe for assigning probabilities to real events. I will argue that the following list of six "laws of probability" is a good representation of our accumulated knowledge related to probabilistic phenomena and that it is a reasonably accurate representation of the actual applications of probability in science.

(L1) Probabilities are numbers between 0 and 1 (including 0 and 1), assigned to events whose outcome may be unknown.

(L2) If events A and B cannot happen at the same time, then the probability that A or B will occur is the sum of the probabilities of the individual events, that is, $P(A \text{ or } B) = P(A) + P(B)$.

(L3) If events A and B are physically independent then they are independent in the mathematical sense, that is, $P(A \text{ and } B) = P(A)P(B)$.

(L4) If events A and B are symmetric, then the two events have equal probabilities, that is, $P(A) = P(B)$.

(L5) When an event A is observed, then the probability of B changes from $P(B)$ to $P(A \text{ and } B)/P(A)$.

(L6) An event has probability 0 if and only if it cannot occur. An event has probability 1 if and only if it must occur.

The laws (L1)-(L6) are implicit in all textbooks, of course. The contents of probability textbooks was selected by a mechanism not unlike Darwinian evolution—the topics most relevant to science established themselves as the core curriculum. The laws (L1)-(L6) are an account of facts and patterns observed in the past—they are a compromise between accuracy, objectivity, brevity, and utility in the description of the past situations involving uncertainty.

9.1 Interpretation of (L1)-(L6)

The laws (L1)-(L6) should be easy to understand for anyone who has any experience with probability; nevertheless, it is a good idea to spell out a few points.

9.1.1 *Events*

An "event" is something that can be determined to have occurred or not. In general philosophy or philosophy of science, the occurrence of an event would be referred to as a "fact." This interpretation of an event applies and is fundamental to not only (L1)-(L6) but also to both frequentist and subjectivist approaches to probability. One must be able to determine whether the relevant event occurred to be able to apply (L6). In the frequency approach, one needs to determine frequencies, so one has to be able to determine which events in a sequence occurred. In the subjectivist approach, the stress is on the Bayes formula. This is a particular case of conditioning so we must be able to tell which events (conditions) occurred.

Sometimes, it is hard to determine whether an event occurred or not. To deal with this problem, we often use random variables instead of events. Consider, for example, the event that tomorrow will be a hot day. The notion of "hot" is quite useful in everyday life but not very useful in science. Instead, scientists use temperature, a physical quantity represented by a "real" number. Tomorrow's temperature is a random variable. On the technical side, random variables can be reduced to events in the following sense. Saying that a random variable X is equal to a is equivalent to saying that events $\{X \leq b\}$ occurred for all $b \geq a$ and they did not occur for any $b < a$.

It is easy to object to the claim that an event must occur or not—life is much more complicated than that. Yes, but an idealization is what we need as long as it is helpful.

Some readers may be surprised that I spend so much time discussing events. The reason is that in the logical philosophy of probability, probability values are assigned to propositions. This idea never gained any traction in the scientific community, and this may be one of the reasons why the logical theory of probability is almost unknown among scientists. Scientists do not have patience and use for the pedantic distinction between the event that an apple fell from a tree, and the proposition "an apple fell from a tree."

9.1.2 *Symmetry*

The word "symmetry" should be understood as the invariance under any transformation preserving the structure of the outcome space (model) and its relation to the outside world. Although I have just attempted to define "symmetry," I doubt that any such definition has any value because the concept of symmetry cannot be reduced to simpler concepts. If someone does not understand the concept of symmetry, there is no reason to believe that he would understand the concept of "invariance" used in my definition.

The mirror symmetry (left and right hands are symmetric in this sense) is an elementary example of non-probabilistic symmetry.

A simple example of (L4) is the assertion that if you toss a coin (once), then the probability of heads is 1/2. The classical definition of probability (see Sec. 2.1) is often applied in situations that involve physical and spatial symmetries, for example, games using dice, playing cards, etc.

From the statistical point of view, the most important examples to which (L4) applies are sequences of i.i.d. (independent, identically distributed) events or random variables, and exchangeable events or random variables (see Chap. 3 for definitions). When observations are ordered chronologically, the i.i.d. property and exchangeability can be thought of as symmetries in time.

Our ability to recognize symmetries relevant to probability is the primary example of resonance in this book. This ability is partly innate, that is, the result of evolution. It is partly taught by example. Without our ability to recognize probabilistic symmetries science would collapse.

It is fundamentally important to realize that (L4) does not refer to the symmetry in a gap in our knowledge, but it refers to the physical (scientific) symmetry in the problem. This point was made as early as 1854 by Boole who thought that our ignorance was not a good justification for assigning equal probabilities to events (see [8], p. 141). We know that the reordering of the results of two tosses of a deformed coin does not affect the probability distribution of the results. But we do not know how the asymmetry of the coin will affect the result of an individual toss. Hence, if we toss a deformed coin twice, and T and H stand for "tails" and "heads," then TH and HT have equal probabilities according to (L4), but TT and HH do not necessarily have the same probabilities.

The above remark about the proper application of (L4) is closely related to the perennial discussion of whether the use of the "uniform" distribution can be justified in situations when we do not have any information. In other words, does the uniform distribution properly formalize the idea of the total lack of information? The short answer is "no." The laws (L1)-(L6) formalize the best practices when some information is available and have nothing to say when there is no information available. I will try to explain this in more detail.

A quantity has the "uniform probability distribution" on $[0, 1]$ if its value is equally likely to be in any interval of the same length, for example, it is equally likely that the quantity is in any of the intervals $(0.1, 0.2)$, $(0.25, 0.35)$ and $(0.85, 0.95)$. Some random quantities can take values in an interval, for example, the percentage of vinegar in a mixture of vinegar and water can take values between 0% and 100%, that is, in the interval [0, 1]. If we have a sample of vinegar solution in water and we do not know how it was prepared, there is no symmetry that would map the percentage of vinegar in the interval (0.25, 0.35) onto the interval (0.85, 0.95). In this case, (L4) does not support the use of the uniform distribution.

If we record the time a phone call is received at an office with the accuracy of 0.1 seconds, then the number of seconds the call was made after the last whole minute is between 0 and 59.9. The time when a phone call is made is close to being stationary (invariant under time shifts) on time intervals of the order of one hour. We can use (L4) to conclude that the number of seconds after the last whole minute when a phone call is received is uniformly distributed between 0 and 59.9.

The uniform distribution can be used as a "seed" in a Bayesian iterative algorithm that generates objectively verifiable predictions. This does not imply that the probabilities described by the uniform prior distribution are objectively true.

9.1.2.1 *The reference class problem and contradictory predictions*

John Smith had a heart attack and smokes cigarettes. What is the probability that he will live at least seventy five years? Suppose that survival data are available for all people who had heart attacks and there are data for people who smoke, but there are no data on people who had heart attacks and smoke. The first data set may yield an estimate of survival probability different from the one generated by the second data set. The (L1)-(L6) theory resolves the problem in a somewhat unsatisfactory way—one cannot find the "correct" probability if significant information about the model is missing. In our example, the information on the combined effect of heart attacks and smoking on survival rates is missing. The "reference class problem" illustrates my point about the applicability of the probability theory. The probability theory provides answers when relevant models and data are available. It does not have an ambition to give rational advice in every uncertain situation.

But not all hope is lost. It has been proved (see [59, 60] and references therein) that predictions based on different data sets cannot be contradictory. More precisely, it is very unlikely that different data sets will generate contradictory predictions, i.e., a prediction (based on one data set) that an event will happen with a very high probability and simultaneously a prediction (based on the other data set) that the event will not happen with a very high probability.

9.1.2.2 *Ergodic theorem*

Applying the concept of symmetry in probability can be considered an instance of the ergodic theorem. See [61, Ch. 6] for the rigorous mathematical presentation of the theorem. From the applied point of view, the ergodic theorem says that the space average equals the time average. For example, the average position of a molecule in a vessel over a long time is the same as the average of positions of all molecules at a fixed time (in the stationary regime).

Consider a die to see how the ergodic theorem can represent an application of symmetry in the probabilistic context. We see that the die is symmetric, and we conclude that the probability of three dots is 1/6 because this is the space average. According to the ergodic theorem, this space average must be equal to the time average, i.e., we predict that the long-range frequency of three dots will be 1/6.

9.1.3 *Enforcement*

The laws (L1)-(L6) are enforced in science and the life of the society. They are enforced not only in the positive but also in the negative sense. A statistician cannot combine two unrelated sets of data, say, on blood pressure and supernova brightness, into one sequence. She has to realize that the combined sequence is not exchangeable, that is, not symmetric. People are required to recognize events with probabilities far from 1 and 0. For example, people are required to recognize that drowning in a swimming pool unprotected by a fence has a non-negligible probability. They have to take suitable precautions; for example, they should fence the pool. Equally simple examples show that people are supposed to recognize both independent and dependent events.

The objectivity of the laws (L1)-(L6) cannot be proved, but the laws can and are enforced. Society assumes and expects that its members can recognize objects and facts, that they know the meaning and value of consequences of actions (money, life, death, suffering), even if they do not personally care about them; that people can follow deductive reasoning and apply the inductive method; that they can act as agents for others, and tasks can be delegated to them. In particular, laws and rules can be enforced. Probability is an inextricable element of this system of beliefs.

9.1.4 *Limits of applicability*

An implicit message in (L1)-(L6) is that there exist situations in which one cannot scientifically assign probabilities. Laws (L1)-(L6) do not say how to assign values to probabilities—they only specify some conditions that the probabilities must satisfy. Only in some cases, such as seven tosses of a symmetric coin, (L1)-(L6) uniquely determine probabilities of all events.

It is quite popular to see the philosophy of probability as an investigation of rational behavior in situations when our knowledge is incomplete. The essence of the science of probability, as embodied in (L1)-(L6), is to present the rules of rational behavior when some information *is* available. In other words, the science of probability is not trying to create something out of nothing, but delineates the boundaries of what can be rationally asserted and verified in situations when some information is available.

The logical, subjective and frequency philosophies and Kolmogorov's axioms contain a misleading implicit message that one has to know probabilities of all events in the sample space to be able to proceed. Knowledge of some restrictions on probability values is enough to generate useful predictions. For example, it is not necessary to know the probability of heads for a deformed coin to be able to generate a prediction that the relative frequencies of heads in two long sequences of tosses of the same deformed coin will be very close to each other. Of course, predictions for deformed coins are not useful, but the same idea applies to truly useful predictions, such as the rate of side effects for a drug.

Laws (L1)-(L6) can be implemented only when we can apply resonance to recognize the relevant properties of events such as symmetry or physical independence. We also need resonance to determine whether an event occurred or not.

9.1.5 *(L1)-(L6) as a starting point*

The relationship between (L1)-(L6) and the real probabilistic and statistical models is analogous to the relationship of, say, Maxwell's equations for electromagnetic fields and a blueprint for a radio antenna. The laws (L1)-(L6) are supposed to be the common denominator for a great variety of methods, but there is no presumption that it should be trivial to derive popular models, such as linear regression in statistics or geometric Brownian motion in finance, from (L1)-(L6) alone. One may find other useful conditions for probabilities besides (L1)-(L6) in some specific situations but none of those extra relations seems to be as fundamental or general as (L1)-(L6).

Some widely used procedures for assigning probabilities are not formalized within (L1)-(L6). These laws are similar to the periodic table of elements in chemistry—a useful and short summary of some basic information, with no ambition for being exhaustive. Consider the following textbook example from frequency statistics. Suppose that we have a sequence of i.i.d. normal random variables with unknown mean and variance equal to 1. What is the best estimator of the unknown mean? This model involves "normal" random variables, so for this mathematical model to be applicable, one has to recognize in practice normal random variables. There are at least two possible approaches to this practical task. First, one can try to determine whether the real data are normally distributed using resonance (common sense or intuition). Alternatively, one can test the data for normality in a formal way, using (L1)-(L6). The last option is more scientific in spirit, but it is not always practical—the amount of available data might be too small to determine convincingly whether the measurements are indeed normal.

A simple model for stock prices provides another example of a practical situation when probabilities are assigned in a way that does not follow directly from (L1)-(L6). A martingale is a process with no overall tendency to go up or down. According to some financial models, stock prices are martingales. The reason is that the stock price is the current best guess of the value of the stock price at some future time, say, at the end of the calendar year. According to the same theory, the current guess has to be the conditional expectation of the future price given the current information. Then a mathematical theorem shows that the stock price has to be a "Doob martingale." Actually, nobody seems to believe that stock prices are martingales. Nevertheless, this oversimplified model makes a prediction that stock prices should be non-differentiable functions of time. The empirical evidence well supports this. My point is that the martingale-like properties of stock prices would be hard to derive from (L1)-(L6) directly.

Finally, let me mention the Schrödinger equation—the basis of quantum mechanics. A solution to the equation can be interpreted as a probability distribution. The laws (L1)-(L6) were involved in the experimental research preceding the formulation of the Schrödinger equation, but I do not see how the probabilities generated by the equation can be derived from (L1)-(L6) alone.

9.1.6 *Statistical models*

Many statistical models and, more generally, probabilistic scientific models are examples of "higher level" resonance. Some simple models are almost direct implementations of (L1)-(L6). Constructing such models requires only the basic level of resonance, that is, skills that are probably innate. Some other models are very complex. The ability to conjecture a good complex model is an acquired skill. The model must be sufficiently realistic to generate reasonably reliable probabilities. It must also be sufficiently simple to be tractable using currently known mathematical methods and computer simulations.

Statistics deals with situations when the model is only partly specified by resonance and logic. Its unspecified parts are represented as unknown "parameters." These can be determined, with some accuracy, from the data.

Traditionally, prior probabilities in Bayesian statistics had a reputation for being subjective, but nobody seemed to have raised a similar objection in the case of models, Bayesian or frequentist. Paradoxically, the resonance theory indicates that models are more "subjective" than (some) prior distributions. Resonance is a mental process which takes sensory data and generates an action or a belief. This process is subconscious, so it cannot be directly verified for correctness no matter how one understands "correctness." Hence, the resonance results' value must be ascertained in another way. One may compare proposals obtained via resonance by different people and use consensus as a substitute for objective truth. One may observe the consequences of actions based on resonance and accept or reject the theory implicit in the actions.

By comparison, many Bayesian priors are "objective" because they require no resonance. For example, consider estimating the probability of heads for a deformed coin ("deformed coin" should be understood as a toy model representing many truly practical situations). A standard Bayesian model for this statistical problem is to assume that the sequence of tosses is exchangeable and, because of de Finetti's theorem, it is a mixture of i.i.d. sequences. The prior distribution of the unknown probability of heads can be taken to be uniform because this gives a very reasonable posterior distribution no matter what the "true" probability of heads is (I put "true" in the quotation marks because only objectivists believe that the probability of heads is an objective number). Note that a group of Bayesian statisticians may agree on using the uniform prior for this particular model under all circumstances, even if they have some personal feelings about the coin. In this case, the prior distribution is fixed in a way that has some scientific justification. But the main point is that the choice of this prior distribution and its justification are fully accessible to conscious analysis, and in this sense, this particular prior distribution, or rather the agreement to use it, is objective.

9.1.7 *The Bayes theorem*

The law (L5) is a form of the Bayes theorem. The reader familiar with the classical version of the Bayes formula (3.1) may be surprised by my claim because (L5) is usually written as $P(B \mid A) = P(A \text{ and } B)/P(A)$ and is considered to be the definition of conditional probability. The last formula is indeed the mathematical definition of conditional probability. But (L5) is a law of science, and it says that the conditional probability is a mathematical quantity that represents real life probability when we acquire new information. The classical Bayes theorem (3.1) is just one of many mathematical ways of calculating conditional probabilities. Of course, (3.1) and its various variants play a crucial role in Bayesian statistics. But I believe that (L5) is the essence of the Bayes theorem in the scientific and philosophical sense.

9.1.8 *Probability of past events*

Laws (L1)-(L6) make no distinction between events that will happen in the future and events that happened in the past.

Probability can be meaningfully attributed to an event if that event can be eventually determined (at least in principle) to have occurred or not. For example, it makes sense to talk about the probability that a scientific theory is correct, the probability that there was life on Mars and the probability that dramas attributed to Shakespeare were indeed written by Shakespeare. In each case, someone can make a prediction, claiming that the probability in question is very close to 0 or 1. In each case, it is conceivable that we will acquire new information at some point that will settle the matter beyond a reasonable doubt. In this way, probabilistic predictions related to these examples can be falsified. Hence, discussing the probability of a scientific theory or a past event makes sense.

9.1.9 *Purely mathematical independence*

Concerning (L3), one should note that there exist pairs of events which are not "physically independent" but independent in the mathematical sense. If you roll a fair die, the event A that the number of dots is even is independent from the event B that the number is a multiple of 3, because $P(A \text{ and } B) = 1/6 = 1/2 \cdot 1/3 = P(A)P(B)$.

9.2 Scientific verification of (L1)-(L6)

A significant difference between probability theory and deterministic branches of science lies in the different method of verifying its assertions.

We assign values to probabilities just like we find a value of any scientific quantity. At the most elementary level, we use resonance directly to determine probabilities. We may use innate resonance or learned resonance. At the scientific level, probabilities are typically determined via a logical process based on information (acquired via resonance). From time to time, specific probability values and probabilistic theories have to be verified to ascertain the reliability of our methods of finding probabilities.

The verification process in probability theory is based on (L6). I will call a claim that gives an event a probability close to 1 a *prediction*. One can scientifically verify probabilistic predictions. One cannot directly verify values of probabilities far from 0 and 1, but there are practical ways to deal with this problem, which will be described later in this section.

The fact that only some probability values (0 and 1) can be directly verified does not diminish the value of the probability theory. In practice, this does not stop scientists from determining values of probabilities far from 0 and 1 and verifying theoretical formulas.

Probability is not the only science that uses verification methods different from those of traditional deterministic sciences. Quantum mechanics is an obvious example of a field with similar epistemological challenges. The difficulties with verification inherent in quantum mechanics are embodied in several laws and have popular-science representations: the Heisenberg uncertainty principle [62], the wave function collapse [63], the double-slit experiment [64], and the Schrödinger cat [65]. And, of course, the probabilistic interpretation of the wave function.

Before I turn to the question of the scientific verification of (L1)-(L6), I will discuss the idea of a scientific proof. The following is a simplified discussion adapted from the book [66] by Lakatos. The idea of a "proof" in mathematics is this: you start with a small set of axioms, and then you use a long chain of logical deductions to arrive at a statement that you consider interesting, elegant or important. Then you say that the statement has been proved. Physicists have a different idea of a "proof"—you start with a large number of unrelated assumptions, you combine them into a single prediction, and you check if the prediction agrees with the observed data. If the agreement is within 20%, you call the assumptions proved.

The procedure for verification of (L1)-(L6) that I advocate resembles the "physics' proof" very much. Consider a real system and assign probabilities to various events using (L1)-(L5), before observing any of these events. Then, use the mathematical theory of probability to find an event A with probability very close to 1 and make a prediction that the event A will occur. The occurrence of A can be treated as a confirmation of the assignment of probabilities, and its non-occurrence can be considered its falsification.

A popular scientific method of verifying probability statements is based on repeated trials—this method is a special case of the general verification procedure described above. It has the same intuitive roots as the frequency theory of probability. Suppose that A is an event, and we want to verify the claim that $P(A) = 0.7$. Then, if practical circumstances allow, we find events $A_1, A_2, \ldots, A_n$ such that n is large and the events $A, A_1, A_2, \ldots, A_n$ are i.i.d. Here, "finding events" means designing an experiment with repeated measurements or finding an opportunity to make repeated observations. Let me emphasize that finding repeated observations cannot be taken for granted. The mathematics of probability says that if $P(A) = 0.7$ and $A, A_1, A_2, \ldots, A_n$ are i.i.d. then the observed relative frequency of the event in the whole sequence will be very close to 70%, with a very high probability. If the observed frequency is indeed close to 70%, this can be considered a proof of *both* assertions: $P(A) = 0.7$ and $A, A_1, A_2, \ldots, A_n$ are i.i.d. Otherwise, one typically concludes that the probability of A is different from 0.7, although in some circumstances, one may instead reject the assumption that $A, A_1, A_2, \ldots, A_n$ are i.i.d.

Recall the discussion of the two interpretations of the "proof," the mathematical and physical ones. Traditionally, the philosophy of probability concerned itself with the verification of probability statements in the spirit of the mathematical proof. One needs to take the physics' attitude when it comes to the verification of probability assignments based on (L1)-(L6), or (L1)-(L6) themselves—it is not only the probability statements but also assumptions about symmetries or lack of physical influence that can be falsified.

The general verification method described above works at (at least) two levels. It is typically used to verify specific probability assignments or relations. However, the combined effect of numerous instances of application of this procedure constitutes a verification of the whole theory, that is, the laws (L1)-(L6).

The method of verification of (L1)-(L6) described above works only in the approximate sense, for practical and fundamental reasons—no events in the universe are absolutely "physically independent," no symmetry is perfect, mathematical calculations usually do not yield interesting events with probabilities exactly equal to 1, and the events of probability "very close" to 1 occur "almost" always, not always.

The actual implementation of experiments or observations designed to verify (L1)-(L6) is superfluous, except for didactic reasons. Scientists have accumulated an enormous amount of data over the centuries, and if someone thinks that the existing data do not provide convincing support for (L1)-(L6), then there is little hope that any additional experiments or observations would make any difference.

9.2.1 *Learning probability*

Verification of (L1)-(L6) is related to learning probability.

In science and everyday life people learn how to apply the laws of nature and also learn how to estimate the initial conditions for a given law in a specific practical situation. Sometimes, the consequences predicted by the law are clearly and immediately observable. However, there are situations when this is not the case, such as if the consequences of actions occur in the distant future. Farmers sow seeds, but they harvest the crops much later. So there is no immediate feedback and verification of the laws governing the growth of plants.

Probability is the phenomenon when one has to apply resonance and knowledge acquired using logic to take action, but its correctness cannot be verified in any single case. One can only verify one's ability to apply resonance and knowledge by collecting many data on the actions of a single person or a community. The Cournot principle can be used to get rid of false beliefs, or one can use the Bayesian approach to learning the correct probabilistic opinions.

Our ability to assign correct probability values to events does not stem only from the individual learning process. Two other sources are instincts acquired in evolution and knowledge obtained from others, spanning the whole spectrum from everyday facts to science.

9.3 Predictions

Law (L6) is a form of Cournot's principle (quoted after [15], p. 585):

> If the probability of an event is sufficiently small, one should act in a way as if this event will not occur at a solitary realization.

The novelty of (L6) lies in its position within the system (L1)-(L6) and my attempts at elucidating its meaning, which are presented below.

I attribute the success of probability theory and statistics to their ability to generate predictions as good as those of deterministic sciences. In practice, no deterministic prediction is certain to occur for various reasons, such as human errors, natural disasters, oversimplified models, limited accuracy of measurements, etc. Some predictions offered by the probability theory in "evidently random" experiments, such as repeated coin tosses, are much more reliable than a typical "deterministic" prediction.

Law (L6) refers to events of probability 0 or 1. In the context of probabilistic phenomena, practically no interesting events have such probabilities—this is almost a tautology. However, there exist many important events whose probabilities are very close to 0 or 1. In other words, the only non-trivial applications of (L6), as an account of the past observations or as a prediction of future events, are approximate. One may wonder whether this undermines the validity of (L6) as a law of science. I believe that (L6) does not pose a philosophical problem any deeper than that posed by the concept of "water." There is no pure water anywhere in nature or any laboratory, and nobody tries to set a universal level of purity for a substance so that it can be called "water"—this is done as needed in scientific and everyday applications.

The concept of temperature applies to human-size bodies, star-size bodies and bacteria-size bodies, but it does not apply to individual atoms. It does not apply to molecules consisting of 3 atoms. The temperature of an atom is not a useful concept, and the same applies to the temperature of a three-atom molecule. How many atoms should a body have so we can talk about its temperature? As far as I know, the critical number of atoms was never defined. Moreover, doing so would not contribute anything to science. Similarly, it would not be helpful to set a number close to 1 and declare that a probabilistic statement is a prediction if and only if it involves a probability greater than that number.

I define a prediction as an event which has probability "very close to 1." This definition may raise some philosophical concerns: (i) a "probabilistic prediction" is not a prediction at all, and (ii) the definition of a prediction is vague, and so "prediction" means different things to different people. These concerns are legitimate, so from the purely philosophical point of view, one could reject the concept of a probabilistic prediction because of (i) and (ii). However, no science has equally high intellectual standards. If we accept the standards implicit in (i) and (ii), social sciences and humanities are almost worthless intellectual endeavors. Moreover, even natural sciences do not offer, in practice, anything more solid than probabilistic predictions (although they do in theory). If we adopted the high standards hinted at in (i) and (ii), then we would have to reject quantum physics, an inherently probabilistic field of science because it would make no predictions at all.

9.3.1 *Predictions at various reliability levels*

The statement of law (L6), "events of probability zero cannot happen," does not distinguish between small but significantly different probability values, say, 0.001 and 10^{-100}. This poses philosophical and practical problems. A person reporting the failure of a prediction to another person does not convey a clear piece of information if the probability of the event was not specified. I believe that a crude rule is needed to make fine distinctions. I will elaborate on this idea using a deterministic example.

Consider the concepts of "black" and "white." The ability to distinguish between black and white is needed in everyday life and science. Not all white objects are equally white; for example, not all white pieces of paper are equally bright. Hence, one could argue that the concept of "white" is insufficiently accurate to be acceptable in science. There are at least two answers to this philosophical problem. The first one is that the crude concept of "white" is sufficiently accurate to be useful.

The second answer is more delicate. One can measure degrees of whiteness, but the measurement process hinges on the human ability (resonance) to distinguish between black and white in a crude way. Imagine a very precise instrument measuring the brightness of "white" paper. The result of the measurement can be displayed using a traditional gauge with a black arrow on the white background or the modern computer screen, using black digits on the white background. Hence, our ability to measure the brightness of the paper with great accuracy depends on our ability to read the gauge or the numbers on the computer screen. This, in turn, depends on our ability to crudely distinguish between white and black. In other words, a fine distinction between degrees of whiteness depends on the crude distinction between white and black.

Measuring physical quantities with great accuracy or measuring extreme values of these quantities is possible, but it usually requires sophisticated scientific theories and superb engineering skills. For example, measuring temperatures within a fraction of a degree from absolute zero, or the (indirect) measurement of the temperature at the center of the Sun require sophisticated theories and equipment. Similarly, to measure very small probabilities with great accuracy, one needs either sophisticated theories or excellent data, or both. In theory, we could use relative frequency to estimate a probability of the order of $10^{-1,000,000}$, but I doubt that we will have the technology to implement this idea soon. The only practical way to determine a tiny probability is first to find a good model for the phenomenon under consideration using observations and statistical analysis and then apply a theorem such as the Large Deviations Principle (see Sec. 3.1.1). The statistical analysis needed in this process involves an application of (L6) in a simple and crude form. For example, one has to reject the possibility that the observed patterns in the data were all created by a faulty computer program. This is effectively saying that an event of a small probability, a computer bug did not happen. Typically, we do not try to determine the order of magnitude of this event in a formal or accurate way—a simple and rough application of (L6) seems to be sufficient and constitutes a part of a very accurate measurement of very small probability.

The above remarks on the relationship between the rough law (L6) and accurate scientific predictions also apply to other elements of the system (L1)-(L6). We must recognize symmetries approximately via simple resonance to apply probability theory and all other scientific theories. Sometimes, this is not sufficient, so scientists measure various physical quantities with great accuracy to determine, among other things, whether various quantities are identical (symmetric). Similarly, applications of probability require that we recognize independent events in a rough way, which is again a case of low-level resonance. In some situations, this is insufficient, and statisticians accurately measure correlation (a number characterizing the degree of dependence).

It is good to keep in mind some examples of events that function in science as predictions. In the context of hypothesis testing (see Sec. 11.3), events with probability 0.95 are treated quite often as predictions. In other words, the "significance level" can be chosen to be 0.05. If an event with probability 0.05 occurs, then this is considered to be a falsification of the underlying theory, that is, the "null hypothesis" is rejected. At the other extreme, we have the following prediction involving a frequency. If we toss a fair coin 10,000 times, the probability that the observed relative frequency of heads will be between 0.47 and 0.53 is about 0.999,999,998. This number and 0.95 are vastly different, so it is no surprise that the concept of prediction is not easy to recognize as a unifying idea for diverse probabilistic and statistical models.

9.3.2 *Predictions in existing scientific and philosophical theories*

Predictions fit naturally into the subjective philosophy and frequency statistics, although in considerably different ways. Suppose a subjectivist wants to make a decision related to an event which has a very high probability. When they calculate the expected value of utility related to a "prediction," that is, the event of very high probability, the expected value will be very close to that if the event had the probability equal to 1. Hence, the decision-maker should make the same decision, no matter whether the probability of the event is very close to 1, or it is exactly equal to 1.

There is a fundamentally important difference, though, between probabilistic and deterministic predictions in the subjective philosophy. Suppose a deterministic event predicted by some theory did not occur in reality. In that case, all people and all theories seem to agree that something must have gone wrong—either the theory was false or its implementation was erroneous. The subjective philosophy does not grant probabilistic predictions (that is, events of high probability) any special philosophical status relative to events with moderate probabilities. If you believe that you will win a lottery with probability

99% and you do not win, the subjective theory has no advice on what you should do, except to stay consistent. You may choose to believe that next time you will win the same lottery with probability 99%.

An important branch of frequency statistics is concerned with hypothesis testing. A hypothesis is rejected if, assuming that the hypothesis is true, the probability that a certain event occurs is very small, and nevertheless, the event does occur. In other words, a frequency statistician makes a prediction, perversely hoping that it will fail, and if indeed the predicted event does not occur, the statistician concludes that the assumptions on which the prediction was based must have been false. The relationship between (L6) and the formal theory of hypothesis testing is similar to that between crude informal measurements and high accuracy scientific measurements described earlier in this section. A sophisticated scientific theory is needed for advanced scientific applications, but it is based on a crude principle (L6) at its foundations.

9.3.3 *Predictions, conditioning and hypothesis tests*

The law (L6) is close in spirit to Fisher's interpretation of hypothesis testing. In his approach, if an event predicted by a model does not occur, we reject the model. Both branches of statistics, frequency and Bayesian, seem to suggest that this method of dealing with unlikely events is too crude. On the frequency side, the Neyman-Pearson theory of hypothesis testing stresses the need for an alternative hypothesis. In other words, we reject the null hypothesis (the theory that makes the prediction) if the observation falls into a certain region defined by both hypotheses, null and alternative. In the Bayesian approach, nothing is ever rejected. An observation of a failed prediction simply modifies our probability distribution, that is, an application of the Bayes theorem generates the posterior distribution from the prior distribution and the observation of the failed prediction.

I believe that the Neyman-Pearson hypothesis testing and the Bayesian approach are meaningful only if we have a good alternative theory, generated by resonance. If the alternative hypothesis is chosen in either theory in an arbitrary way then there is no reason to think that the Neyman-Pearson or Bayesian methods will yield a result that is useful.

9.3.4 *Prediction examples*

I will present a few examples of predictions. I will start by discussing long-run frequencies because this is undoubtedly the most fundamental example of a real-life application of a prediction. I have already mentioned that if we toss a fair coin 10,000 times, then the probability that the observed relative frequency of heads will be between 0.47 and 0.53 is about 0.999,999,998. The close relationship between frequencies and probabilities is so profoundly embedded in the scientific psyche that many people fail to notice that this relationship is just one of many practical cases of predictions.

In many situations, what matters is not the long-run average but a catastrophically large value of a process. The process may be i.i.d. or have i.i.d. increments, or it may have a completely different structure. Examples of catastrophically large values of random variables include intensive rainfall resulting in flooding and high-velocity wind during hurricanes. Both natural phenomena cause huge losses. In this context, predictions say that catastrophically large values of the relevant quantities are improbable.

In other cases, what matters is not a particularly large value of a random variable but a complex catastrophic event such as a nuclear plant meltdown. This event may be related to unusually large values of some quantities, such as the level of radiation in the air, but reducing the catastrophe to these large values seems to be missing the point. A prediction related to such an event is a claim that the catastrophe has a tiny probability.

Some of the most significant and practical examples of

probabilistic predictions are never represented as such in everyday life. Everyone knows an egg will never unscramble itself, and water on a hot stove will not freeze. Nevertheless, these events are not impossible—according to modern physics, they are merely highly improbable. Two sources of randomness at the microscopic level generate deterministic phenomena on the macroscopic level—statistical physics and quantum physics. The egg and water examples illustrate some statistical physics effects. Tunneling diodes are quantum mechanical devices in which electrons jump ("tunnel") over a seemingly insurmountable barrier. On a macroscopic scale, a cat could "tunnel" through a door, but this event is extremely unlikely to occur, so people install cat doors.

9.3.5 *Surprise*

In July 2022, it was reported that two FBI officials had been chosen for extremely rare tax examinations. The IRS was suggested to have targeted the two officials because they were political opponents of the US President. The IRS asserted that the two individuals had been chosen randomly.

The probability that given two individuals, both of them will be chosen randomly for an audit is very small—about one in 950 million. Since the event occurred, it appears that (L6) suggests that the two officials were not chosen randomly.

The problem with this reasoning is that the observed event (the audit of the two specific officials) was not specified in advance. It is fair to say that people in general and especially mass media are looking for surprising and, therefore, interesting events. The probability of finding an "unusual" event cannot be calculated in a simple way because it depends on social patterns. I do not think that it is impossible to estimate the probability of finding an "unusual" event, but this would require constructing a convincing social model, collecting many data and analyzing them carefully.

Personally, I am convinced that the two officials were indeed targeted for political reasons. However, I do not know how to build a realistic probabilistic model supporting my intuition.

The book by [67] titled *The Black Swan* is dedicated to analyzing the impact of surprising events in real life.

9.3.6 *Histograms and image reconstruction*

The rise of computers and their ability to generate pictures representing data illustrates predictions in a far more interesting way than the classical predictions of stable frequencies. A popular example of graphical data representation is a histogram of the empirical distribution (more precisely, its density). The shape of the graph is a prediction. For example, if the empirical distribution comes from the normal distribution, the graph will have the characteristic bell shape with high probability. A more exotic example is fractional Brownian motion. Figures 14.2–14.4 show simulations of fractional Brownian motion with three different values of the "Hurst parameter." The graph becomes smoother as the parameter becomes larger. One look at the graphs in Figures 14.2 and 14.4 is all one needs to tell which of the two processes has a larger Hurst parameter. The roughness of the graph is a prediction—it occurs with very high probability, not certainty.

Sometimes, photographs are transmitted through noisy channels, for example, when sent from space. Different techniques of reconstruction make the photographs look much better to the human eye; see an example in Fig. 9.1. The prediction here is that the reconstructed photograph is very close to the true representation of the object with very high probability. The idea that the reconstructed image is "very close" to the original can be expressed in rigorous mathematical terms. In practice, it means that the objects in the reconstructed photograph are easily recognizable by humans.

Fig. 9.1 The picture on the left is a photograph corrupted by placing 10,000 white pixels in a random (i.i.d. uniform) fashion. The corrupted photograph was reconstructed using the "MedianFilter" Mathematica function. The result is on the right.

9.3.7 *Multiple predictions*

When it comes to deterministic predictions, all predictions are supposed to hold simultaneously, so the failure of a single prediction may falsify a whole theory. By nature, probabilistic predictions may fail even if the underlying theory is correct. If a large number of predictions are made at the same time, then it is possible that, with high probability, at least one of the predictions will fail. This seems to undermine the idea that a failed prediction falsifies the underlying theory. In practice, the problem is dealt with using (at least) three processes that I will call *selection*, *aggregation* and *amplification*. Before I explain these concepts in more detail, I will describe a purely mathematical approach to the problem of multiple predictions.

Consider independent predictions $A_1, A_2, A_3, \dots$. Suppose that all these events have the probability 99%. Then, in the long run, these predictions will fail at an approximate rate of 1%, even if the theory behind these predictions is correct. The last statement is a prediction itself and can be formalized as follows, using the Law of Large Numbers. There exists a large number n_0 such that for every number n greater than n_0, the percentage of predictions A_k that fail among the first n predictions will be between 0.9% and 1.1%, with probability greater than 99%. The last probability can be increased, but then the value of n_0 must be adjusted.

The process of *selection* of predictions is applied mostly subconsciously. We are surrounded by a very complex universe, full of unpredictable events. Most of them are irrelevant to our lives, such as whether the leaf at the top of my linden tree will fall to the north or the south. We normally think about a small selection of events that can influence our lives, and we try to make predictions concerning these events. The fewer the number of predictions, the fewer the number of failed predictions.

A lottery provides an example of the *aggregation* procedure. Typically, the probability that a specific person will win a given lottery is minuscule. In other words, we can make a prediction that the person will not win the lottery. The same prediction applies to every person playing the same lottery. However, it is not true that the probability that nobody will win the lottery is small. We never make one hundred thousand predictions, each one saying that a different specific person will not win the lottery. The number of predictions that are made is reduced by combining large families of related predictions into a smaller number of "aggregate" predictions. For some lotteries, one can make a single "aggregate" prediction that somebody will win the lottery. For some other lotteries, the probability that someone will win the lottery may be far from 1 and 0—in such a case, no aggregate prediction can be made.

When a specific prediction is very significant to us, we can *amplify* its power by collecting more data. This is a standard practice in science. For example, most people believe that smoking increases the probability of cancer. Let me represent this claim in a somewhat artificial way as a statement that the cancer rate among smokers will be higher than the cancer rate among non-smokers in 2035 with probability p. We believe that p is very close to 1, but to make this prediction even stronger, new data on smokers and cancer victims are continually collected. The more data are available, the higher the value of p they justify. Both amplification and aggregation are used in the context of hypothesis testing (see Sec. 11.3.2).

Even after applying selection and aggregation, a single person will generate many predictions over his lifetime. If the predictions are reasonably independent then one can prove, just like indicated earlier in this section, that only a small proportion of predictions will fail, with high probability. This claim is a single "aggregate" prediction. Such a single aggregate prediction can be constructed from all predictions made by a single physical person or by a group of people, for example, scientists working in a specific field.

I will now address a few more scientific and philosophical points related to multiple predictions.

First, it is interesting how mathematicians approach extensive families of predictions. Consider, for example, Brownian motion. This stochastic process is a mathematical model for a chaotically moving particle. Let $B(t)$ denote the position of a Brownian particle moving along a straight line at time t. It is known that for a fixed time t, with probability one, $B(t)$ is not equal to 0. It is also known that for a fixed time t, with probability one, the trajectory $B(t)$ has no derivative; that is, it is impossible to determine the velocity of the Brownian particle at time t. The number of times t is infinite and, moreover, it is "uncountable." Can we make a single prediction that all of the above predictions about the Brownian particle at different times t will hold simultaneously? It turns out that, with probability one, for all times t simultaneously, there is no derivative of $B(t)$. However, with probability one, there are times t such that $B(t)$ equals 0. These examples show that infinitely many (uncountably many) predictions can be combined into a single prediction or not, depending on the specific problem. On the technical side, this is related to the fact that the product of zero and infinity is not a well-defined number. In practice, nobody makes infinitely many separate predictions about $B(t)$ for all values of t. One can make either several simultaneous predictions for a few specific values of t, or an aggregate prediction concerning the behavior of the whole Brownian trajectory.

Scientists make large numbers of diverse predictions, and the same holds for ordinary people, except that in the latter case, the predictions are informal. Transforming a family of predictions into a single aggregate prediction may be hard for several reasons. First, using mathematics to combine multiple predictions into a single aggregate prediction may be easier said than done when individual predictions are not independent. Second, the single combined prediction can be verified only at the end of a possibly long period. Third, on the philosophical side, a single combined prediction is not an attractive idea either. The falsification of a single aggregate prediction only indicates that there is something wrong with the theory underlying all of the constituent predictions. A single falsified aggregate prediction provides little specific information on what might have gone wrong because it is based on a very complex theoretical structure. Rather than combine multiple predictions into a single aggregate prediction, a more practical way to go is to treat an individual falsified probabilistic prediction not as a proof that the underlying theory is wrong but as an indication that it may be wrong, and hence it merits further investigation. The amplification procedure described above, that is, collecting more data, can generate a new prediction with probability very close to 1. A large number of predictions will hold simultaneously with a very high probability if the constituent predictions hold with even greater probability.

The resonance theory suggests that we believe that predictions will come true because we were shaped to hold such beliefs by the evolution. Animals perceive probabilities in an instinctive way. The selection process made animals, including humans, believe in positive outcomes of highly probable events because such beliefs increased the survival probability.

9.4 Symmetry, independence and resonance

Our ability to recognize symmetries and independence is a special case of resonance. The success of probability theory (that is, good predictions in the sense of (L6)) stems from reliable information about some aspects of the real world. My theory asserts implicitly that symmetries and physical independence are objective and can be effectively recognized. If one adopts the philosophical position that objective symmetries do not exist or cannot be effectively recognized, then the probability theory becomes practically useless. I have shown in Chap. 5 and Chap. 6 the frequency and subjective theories are meaningless without (L3) and (L4). Hence, if there is a genuine problem with these two laws, all philosophies of probability are severely affected.

The problem is not unique to the probability theory and (L1)-(L6). Let me reiterate some claims from Chap. 7 and Chap. 8, this time specialized to symmetry and independence. The ability to recognize events which are symmetric or physically unrelated is a fundamental element of any scientific activity. The need for this ability is so basic and self-evident that scientists rarely talk about it. Suppose a biologist wants to determine whether zebras are omnivorous. He has to go to Africa and observe a herd of zebras. He has to find a collection of symmetric objects characterized by black and white stripes. In particular, he must not mistake lions for zebras. Moreover, the zoologist must disregard any information that is unrelated to zebras, such as data on snowstorms in Siberia in the seventeenth century or car accidents in Brazil in the last decade. Skills needed to recognize symmetry in the probabilistic context are precisely the same as those needed to count objects. People are expected and required to recognize symmetries; for example, shoppers and sellers are expected by society to agree on the number of apples in a basket—otherwise, commerce would cease to exist.

Every scientific experiment in deterministic science is based on symmetry and independence. Symmetry may be exemplified by the identity of repeated trials, or it may be the identity of the model and a single experiment. The concept of independence is implicit in attempts to isolate the experiment from all noises and outside influences.

The laws (L1)-(L6) are based on principles taken for granted elsewhere in science, if not in philosophy. One cannot prove beyond any doubt that people can effectively recognize events that are disjoint, symmetric or physically independent. However, probability theory cannot be implemented successfully if people cannot do that.

9.5 Applications of (L1)-(L6): Some examples

Anyone who has ever had any contact with science and its applications knows that (L1)-(L6) are a *de facto* scientific standard, just like Newton's laws of motion.

The laws (L3) and (L4) are used in probability in the same way as in science. Recall the example involving zebras in Sec. 9.4. When a scientist wants to study some probabilistic phenomena, he often finds collections of symmetric objects. For example, if a doctor wants to study coronary heart disease, he has to identify people, as opposed to animals, plants and rocks. This is considered so obvious that it is never mentioned in science. More realistically, physicians often study some human subpopulations, such as white females. This is a little more problematic because the definition of race is unclear. Doctors apply (L3) by ignoring data on volcano eruptions on other planets and observations of Moon eclipses.

9.5.1 *Poisson process*

A non-trivial illustration of (L4) is a "Poisson process," a popular model for random phenomena ranging from radioactive decay to telephone calls. This model is applied if we know that the number of "arrivals" (for example, nuclear decays or telephone calls) in a given interval of time is independent of the number of arrivals in any other (disjoint) interval of time. The model can be applied only if we assume in addition a symmetry, specifically the "invariance under time shifts"—the number of arrivals in a time interval can depend on the length of the interval but not on its starting time. It can be proved in a rigorous way that the independence and symmetry described above uniquely determine a process, called a "Poisson process," except for its intensity, that is, the average number of arrivals in a unit amount of time.

9.5.2 *Laws (L1)-(L6) as a basis for statistics*

Laws (L1)-(L6) are applied by all statisticians, frequency and Bayesian. A typical statistical analysis starts with a "model," that is, a set of assumptions based on (L1)-(L6). Here (L2), (L3) and (L4) are the most relevant laws, as in the example with the Poisson process. The laws (L1)-(L6) usually do not specify all probabilities or relations between probabilities, such as the intensity in the case of the Poisson process. Frequency statisticians consider the intensity to be an "unknown but fixed" parameter that must be estimated using available data. Bayesian statisticians treat the unknown parameter as a random variable and give it a distribution known as a *prior*. I will discuss the frequency and Bayesian branches of statistics in much more detail later in the book. I want to stress that (L1)-(L6) are used by both frequency and Bayesian statisticians. The application of these laws, especially (L3) and (L4), has nothing to do with the official philosophies adopted by the two branches of statistics. Frequency statisticians apply (L3) and (L4) even if the available samples are small. Models (but not all priors) used by Bayesian statisticians *de facto* follow the guidelines given in (L1)-(L6), and so they attract very little philosophical controversy.

9.5.3 *Long run frequencies and (L1)-(L6)*

I will show how the long run frequency interpretation of probability fits into the framework of (L1)-(L6). I have already criticized, in Chap. 5, the von Mises theory of "collectives" formalizing the idea of the long-run frequency. I will now argue that this staple scientific application of probability agrees well with (L1)-(L6).

Concretely, consider a clinical trial of a new drug. For simplicity, assume that the result of the test can be classified as a "success" or "failure" for each individual patient. Suppose that you have a "large" number n of patients participating in the trial. There is an implicit other group of patients of size m, comprised of all people afflicted by the same malady in the general population. We apply the law (L4) to conclude that all $n + m$ patients form an "exchangeable" sequence. Choose an arbitrarily small number $\delta > 0$ describing your error tolerance and a probability p, arbitrarily close to 1, describing the level of confidence you desire. One can prove that for any values of $\delta > 0$ and $p < 1$, one can find n_0 and m_0 such that for $n > n_0$ and $m > m_0$, the absolute value of the difference between the success rate of the drug among the patients in the clinical trials and the success rate in the general population will be smaller than δ with probability greater than p. One usually assumes that the general population is large and so, implicitly, $m > m_0$, whatever the value of m_0 might be. If the number of patients in the clinical trial is sufficiently large, that is, $n > n_0$, one can apply (L6) to treat the clinical trial results as a predictor of the future success rate of the drug.

In other applications of the idea of the long run frequency, the counterpart of the group of patients in the clinical trial may be a sequence of identical measurements of an unknown constant. In such a case, the general population of patients has no explicit counterpart—this role is played by all future potential applications of the constant.

9.5.4 *Life on Mars*

The question of the probability that there was or there is life on Mars is of great practical importance. If the reader is surprised by my claim, they should think about the enormous amount of money—billions of dollars—spent by several nations over many years on spacecraft designed to search for a sign of (past or present) life on the surface of Mars. A good way to present the probabilistic and philosophical challenge related to life on Mars is to pose the following question: Why is it rational to send life-seeking robots to Mars but not to Venus?

The frequency theory of probability suggests that we should look for a long sequence of "identical" events that incorporates life on Mars, and another sequence for Venus. A natural idea would be to look at a long sequence of planets similar to Mars and see what percentage of them ever supported life. There are several problems with this idea. The first is somewhat philosophical but cannot be ignored from the scientific perspective. Which planets are "similar" to Mars? Should we consider all planets that have the same size and the same distance from their star? Or should we insist that they also have a similar atmosphere and a similar chemistry of the rocks on the surface? If we specify too many similarities, then our "sequence" will consist of a single planet in the universe—Mars. A much more practical problem is that at this time, we cannot observe even a small sample of planets "similar" to Mars and verify whether they support life. Even if the "long run frequency of life on planets similar to Mars" is a well-defined concept, it is of no help to scientists and politicians trying to decide whether they should spend money on life-seeking spacecraft.

The subjective theory does not offer much practical advice either. This theory stresses the need to be consistent. Trivially, it is consistent to believe that there was life on Mars but not on Venus with high probability, and it is also consistent to believe that there was life on Venus but not on Mars with high probability.

Laws (L1)-(L6) can be used to justify not sending life-seeking spacecraft to Venus in the following way. Multiple observations of and experiments with different life forms on the Earth show that life known on the Earth can survive only in a certain range of conditions. The spectrum of environments that can support life is enormous, from ocean depths to deserts, but there seem to be limits. The environment on Venus is more or less similar to ("symmetric with") some environments created artificially in laboratories. Since no life survived in laboratory experiments in similarly harsh conditions, we believe that life did not and does not exist on Venus. We use an approximate symmetry to make a prediction that there was no life on Venus. The argument uses laws (L4) and (L6).

The question of life on Mars illustrates well the "negative" use of (L6). What we seem to know about Mars suggests that there might have been times in the past when the environment on Mars was similar to ("symmetric with") environments known on the Earth or created artificially in laboratories where life was sustained. Hence, we cannot conclude that the probability of life on Mars is very small. That does not mean that the probability is large. The only thing that we can say is that we cannot make the prediction that signs of life on Mars will never be found. Hence, sending a life-seeking spaceship to Mars is not irrational. It is not irrational to stop sending life-seeking spaceship to Mars either. In a situation when neither an event nor its complement has a very small probability, no action can be ruled out as irrational, and the decision is a truly subjective choice.

The last assertion needs clarification. Decision-makers often attach significance not to the outcome of a single observation or experiment but only to the aggregate of these. For example, stores are typically not interested in the profit made on a single transaction but in the aggregate profit over some time, say a year. A decision-maker has to choose an aggregate of decisions that is significant for him. One could argue that one should consider the biggest aggregate possible, but that would mean that we would have to consider all our personal decisions in our lifetime as a single decision problem. This may be the theoretically optimal decision strategy, but it is hardly practical. Thus, most decision-makers consider "natural" aggregates of decisions. An aggregate may consist of a single decision with significant consequences. My suggestion that sending life-seeking space probes to Mars and not sending them are both rational decisions, was made under the assumption that this action was considered in isolation. Politicians are likely to consider many spending decisions as an aggregate, so one could try to make a prediction about the cumulative effect of all such decisions. The decision to send spacecraft to Mars may be a part of an aggregate of decisions that is rational or irrational as an aggregate, without specifying the rationality status of the individual decisions in the aggregate.

9.5.5 *Black ravens*

A well-known paradox is concerned with black ravens. Let H be the claim that all ravens are black. Then, the claim H' that "all non-black objects are not ravens" is logically equivalent to H. If we observe a black raven, this evidence seems to support H. By the same logic, if we observe a white piece of paper, this piece of evidence supports H'. Since H and H' are logically equivalent, observing a white piece of paper supports the claim that all ravens are black. This sounds implausible.

The laws (L1)-(L6) suggest a very simple and natural resolution of the paradox. We have to realize that the coloring of ravens is independent of the coloring of objects different from ravens—this is an application of resonance in the specific form of (L3). In other words, natural processes that generate the coloring of ravens are independent in the physical sense from natural and human-controlled processes that generate the coloring of objects that are different from ravens. Hence, the probability that the next observed raven is black is independent of the observed colors of non-ravens.

The argument is oversimplified, of course, because ravens are related to crows, so observations of crows could yield information about the coloring of ravens. While this makes the problem more complicated, it does not invalidate the main point—an application of (L3), even if in an approximate form, solves the apparent paradox.

9.6 Beauty and probability are communicable

Beauty is communicable. We all have our own opinions about what is beautiful and what is not, but we can effectively and reliably learn what other people consider beautiful. Both beauty and probability are communicable.

Beauty is communicable in the following sense.

(i) We can learn what some other individual feels about an object, person, book, etc., and we can relate this knowledge to our own feelings about the same or different object, person, etc.
(ii) We can determine statistical trends about beauty, elegance, etc. They are too chaotic and unpredictable to be the basis of a scientific theory or a measurement method, but knowing the statistics gives objective information that can be used in practice, for example, by businesses selling consumer goods.

Probability is communicable in that we can effectively and reliably learn other individuals' opinions about events with very small probabilities, physical independence, frequency-based probability estimates, and occurrence of events. We can also discern statistical trends in probabilistic opinions.

When a scientist says that the probability of a side effect for a particular drug is 7%, then all people, objectivists and subjectivists alike know what this means objectively—the scientist did repeated experiments, recorded the relative frequency of side effects and announced that the probability is the same as the relative frequency, 7%. All of this is objective in the sense that we humans have the ability to communicate about probability, just like we can communicate about beauty, electricity, arithmetic and miracles. This does not mean that the person who receives information about side effects has to believe in it or its consequences. In particular, the recipient of the information does not have to identify probability and frequency.

While people's opinions on beauty, religion, elegance, politics, etc., vary widely, experts' opinions are almost identical on mathematical and scientific subjects. Probabilistic views are almost identical in many practical situations. This, when combined with communicability, makes probability as objective as any other science from the operational point of view.

9.7 Man made probability in his image

According to [68], "The Image of God is a concept and theological doctrine in Judaism, Christianity, and some Sufism sects of Islam, which asserts that human beings are created in the image and likeness of God." Ludwig Feuerbach, in his *Lectures on the Essence of Religion*, put the idea on its head, asserting that "God did not, as the Bible says, make man in His image; on the contrary man [...] made God in his image" (see [69] for a detailed discussion).

It is widely believed that "history is written by the victors," although the claim is controversial.

The two examples above illustrate the same idea—theoretical ideas and writings may be the mere reflection of preexisting human opinions rather than the representation of reality or truth. It is natural to regard this as an imperfection of research in fields outside empirical sciences. My laws (L1)-(L6) openly follow this pattern—I am trying to codify what people already believe in. I propose two justifications for my approach. First, the patterns of human thinking about randomness should be codified so that students can learn what will be expected of them if they use probability and statistics at work.

Second, one could argue that our probabilistic instincts have been shaped and tested in evolution (see Sec. 14.5.7), so they are empirically confirmed, at least partially. I realize that the second argument is relatively weak. I reject it in the cases of Catholic theology and history writings. Also, psychological research shows limitations of and problems with our probabilistic instincts—see [70]. Nevertheless, I believe that the probabilistic behavior of our animal ancestors had to be quite practical to enhance their chances of survival. Social animals (for example, ants and bees) display behaviors that are not beneficial for the individual but are beneficial for the community. So we, as outsiders, should be able to discern "rational" patterns in the behavior of social animals—I believe that they agree with (L1)-(L6).

Chapter 10

Decision Making

Decision making is not a part of science (see Sec. 8.4). Science can (try to) predict the consequences of various decisions, but it is not the role of science to tell people what they should do.

I will divide my discussion of decision making into several parts. Section 10.2 will deal with decision-making options when all the relevant probabilities are determined by (L1)-(L6) or in some other way. Section 10.4 will address the question of what to do when not all probabilities needed to make a decision are known.

10.1 Common practices

Some common beliefs and practices concerning decision making involving uncertainty seem to be almost uncontroversial. These beliefs inspire various scientific and philosophical proposals for formal decision-making theories. Decision-making principles pose a problem when a situation does not fall squarely into one of the following two categories.

If an event of interest has a very high probability, then it is natural to assume that it will happen with certainty. This common belief is the essence of my law (L6). People typically ignore events of small probability, such as the possibility of dying in a car accident on a given trip. Probabilistic decision making is thus reduced to deterministic decision making in some situations. One could even go as far as to say that all "deterministic" decision problems arise in this way because nothing is ever totally certain.

The other type of decision problem which does not seem to pose much of a practical or intellectual challenge is when the decision-maker is faced with a long sequence of similar decisions. Then it is natural to maximize the expected value of the gain or the expected value of the utility in each case and hope that the Law of Large Numbers will generate the maximum average gain over a long time. The details of what should be maximized, what probabilities should be used, etc., are the subject of a philosophical and practical debate, but it is clear that most people are willing to choose individual decisions based on long-run average predictions.

I will return to this discussion in Sec. 10.2.5.

10.2 Decision making in the context of (L1)-(L6)

I will start with a semi-formal description of a simple probabilistic decision problem. Few real-life decision problems are that simple, but readers unfamiliar with formal decision theory might get a taste of it. Suppose one has to choose between two decisions, D_1 and D_2. Suppose that if decision D_1 is made, the gain may take two values G_{11} and G_{12}, with probabilities p_{11} and p_{12}. Similarly, D_2 may result in rewards G_{21} and G_{22}, with probabilities p_{21} and p_{22}. Assume that $p_{11} + p_{12} = 1$ and $p_{21} + p_{22} = 1$, all four probabilities are strictly between 0 and 1, and $G_{11} < G_{21} < G_{22} < G_{12}$ so that there is no obvious reason why D_1 should be preferable to D_2 or vice versa. Recall that, in this section, I assume that the four probabilities, p_{11}, p_{12}, p_{21} and p_{22}, are determined by (L1)-(L6) or in some other way. How can we decide in a "rational" way which of the two decisions D_1 or D_2 is "better"? I will start by criticizing the most popular answer to this question, and then I will propose two other decision-making philosophies.

10.2.1 *Maximization of expected gain*

A standard decision-making philosophy is to choose a decision that maximizes the expected gain. This decision-making philosophy is quite intuitive, but I will show it is profoundly flawed.

Recall the decision problem from the previous section. If we make decision D_1 then the expected gain is $G_{11}p_{11}+G_{12}p_{12}$ and if we make decision D_2 then the expected gain is $G_{21}p_{21}+G_{22}p_{22}$. Hence, if we want to maximize the expected gain, we should make decision D_1 if $G_{11}p_{11}+G_{12}p_{12} > G_{21}p_{21}+G_{22}p_{22}$, and we should choose D_2 if the inequality goes the other way. The decisions are equally preferable if the expected values are equal.

The above strategy sounds rational until we recall that, typically, the "expected value" is not expected at all. If we roll a fair die, the "expected number" of dots is 3.5. Of course, we do not expect to see 3.5 dots. I feel that most scientists subconsciously ignore this simple lesson. To emphasize the true nature of the "expected value," let me use an equivalent but much less suggestive term "first moment." "Maximizing the first moment of the gain" sounds much less attractive than "maximizing the expected value of the gain." Why should one try to maximize the first moment of the gain and not minimize the third moment of the gain? I will address this question from both frequency and subjective points of view.

The frequency theory of probability identifies the probability of an event with the limit of relative frequencies of the event in an infinite sequence of identical trials, that is, a collective. Similarly, the expected value (first moment) of a random variable may be identified with the limit of averages in an infinite sequence of i.i.d. random variables, by the Law of Large Numbers. Suppose we want to use the frequency theory to justify maximizing the first moment of the gain. In that case, we have to assume that we face a long sequence of independent and identical decision problems and the same decision is made every time. Only in some practical situations does one decision-maker deal with a sequence of independent and identical decision problems.

In everyday life, various decision problems may have a completely different structure. In science and business, the form of a decision problem may sometimes remain the same. Still, the information gained in the course of analyzing earlier problems may be applied in later problems, and so the decision problems may not be independent. The frequency theory of probability provides a direct justification for the practice of maximizing the expected gain only in some cases.

Maximizing the expected gain within the subjective theory of probability seems to be a reasonable strategy for the same reason as in the case of the frequency theory—linguistic. The subjective theory says that the only goal that a decision-maker can achieve is the elimination of all Dutch books by choosing a consistent decision strategy. There are countless ways in which one can achieve consistency, and none of them is any better than any other in any objective sense, according to the subjective theory. A mathematical theorem says that if you choose a consistent strategy, then you maximize the expected gain, according to some probability distribution. The idea of "maximizing of expected gain" clearly exploits subconscious associations of decision-makers. They think that their gain will be large if they choose a decision that maximizes the expected gain. The subjective theory says that the gain can be large or small (within the range of possible gains corresponding to a given decision), but one cannot prove in any objective sense that the gain will be large. Moreover, the subjective theory teaches that when the gain is realized, its size cannot prove or disprove in the objective sense any claim about the optimality or suboptimality of the decision made. Hence, maximizing the expected gain means maximizing the subjective feelings about the gain. This sounds like a piece of advice from a "self-help" book rather than science.

I will present the above claims in a somewhat different way. Within the subjective philosophy, the idea of maximizing of subjective gain is tautological. The prior distribution can be presented in various formal ways. One of them is to represent the prior as a set of beliefs containing (among other statements) conditional statements of the form "if the data turn out to be x then my preferred decision will be $D(x)$." Since, in the subjective theory, probabilities and expectations are only a way of encoding consistent human preferences, an equivalent form of this statement is "given the data x, the decision $D(x)$ maximizes the expected gain." Hence, the question of why you would like to maximize the expected gain is equivalent to the question of why you think that the prior distribution is what it is. In the subjective philosophy, it is not true that you should choose the decision which maximizes the expected gain; the decision that maximizes the expected gain was labeled as such because you said you preferred it over all other decisions. The strict subjectivist version of the Bayesian ideology successfully obfuscates the circularity of the subjectivist preference for the maximization of the expected gain. A Bayesian subjectivist starts with a prior distribution (prior opinion), then collects the data, combines the prior distribution and the data to derive the posterior distribution, and finally makes a decision that maximizes the expected gain, according to the posterior distribution. The whole multistage process, often very complex from the mathematical point of view is a smokescreen that hides the fact that the maximization of the expected gain according to the posterior distribution is nothing but the execution of the original (prior) preference, not assumed by subjectivists to have any objective value.

10.2.2 *Maximization of expected gain as an axiom*

Before I propose my own two alternative decision-making philosophies, I have to mention an obvious, but repugnant to me, philosophical choice—one can adopt the maximization of the expected gain as an axiom. I have argued in Sec. 8.4 that the choice of a decision strategy is not a part of the science of probability, so this axiom cannot be shown to be objectively correct or incorrect, except in some special situations. Hence, I am grudgingly willing to accept this choice of the decision philosophy if anyone wants to make this choice. At the same time,

I firmly believe that the choice is based on a linguistic illusion. If the same axiom were phrased as "one should maximize the first moment of the gain," most people would demand a good explanation for such a choice. I have already shown that the justifications given by the frequency and subjective theories are unconvincing.

The honest answer to the question "Why is it a good idea to maximize the expected gain?" seems more technical than philosophical in nature. An excellent technical reason to use expectation is that it is additive, that is, the expectation of the sum of two random variables is the sum of their expectations, no matter how dependent the random variables are. This is very convenient in many mathematical arguments. The second reason is that assigning a single value to each decision makes all decisions comparable, so one can always find the "best" decision. Finding the "optimal" decision is often an illusion based on clever manipulation of language, but many people demand answers, even poor answers, no matter what.

The maximization of the expected gain can be justified, at least in a limited way, within each of the two decision-making philosophies proposed below. I find that approach much more palatable than the outright adoption of the expected gain maximization as an axiom.

10.2.3 *Stochastic ordering of decisions*

The first of my proposals for a decision philosophy is based on the idea that probability is the only quantity that distinguishes various events within the probability theory. I will use an analogy to clarify this point. Consider two samples of sulfur, one spherical and one cubic. If they have the same mass, they are indistinguishable from the point of view of chemistry. Similarly, two balls made of different materials but with the same radii and the same density would be indistinguishable from the point of view of the gravitation theory. Consider two games, one involving a fair coin and the other involving a fair die. Suppose that you can win \$1 if the coin toss results in heads and lose \$2 otherwise. You can win \$1 if the number of dots on the die is even, and otherwise, you lose \$2. Since the probabilities are the only quantities that matter in this situation, there is no rational reason to prefer one of the games over the other.

Now consider two games whose payoffs are known and suppose that they are stochastically ordered, that is, their payoffs G_1 and G_2 satisfy $P(G_1 \geq x) \geq P(G_2 \geq x)$ for all x. It is elementary to see that there exist two other games with payoffs H_1 and H_2 such that G_k has the same distribution as H_k for $k = 1, 2$, and $P(H_1 \geq H_2) = 1$. The game with payoff H_1 is obviously more desirable than the one with payoff H_2, and, therefore, the game with payoff G_1 is more desirable than the one with payoff G_2, by the equivalence described in the previous paragraph. In other words, the decision-making philosophy proposed here says that a decision is preferable to another decision if and only if its payoff stochastically majorizes the payoff of the other decision.

Here are some properties of the proposed decision-making strategy.

(i) Consider two decisions and suppose that each one can result in a gain of either a or b dollars. Then the gain distributions are comparable. In this simple case, the proposed decision algorithm agrees with maximizing the expected gain.

(ii) Two decisions may be comparable even if their expected gains are infinite (that is, equal to plus or minus infinity), or undefined.

(iii) If two decisions are comparable and the associated gains have finite expectations, a decision is preferable to another decision if and only if the associated expected gain is larger than the analogous quantity for the other decision.

(iv) Suppose that in a decision problem, two decisions D_1 and D_2 are comparable and D_1 is preferable. Consider another decision problem consisting of comparable decisions D_3 and D_4, with D_3 being preferable. Assume that all random events involved in the first decision problem are independent of all events involved in the second problem. If we consider an aggregate decision problem in which we have to make two choices, one between D_1 and D_2, and another choice between D_3 and D_4, then the aggregate decision D_1 and D_3 is comparable to the aggregate decision D_2 and D_4, and the first one is preferable. Unfortunately, the same conclusion need not hold without the assumption of independence of the two decision problems.

(v) One can justify expected gain maximization (under some circumstances) using the idea of stochastic ordering of decisions as follows. Suppose that one has to deal with n independent decision problems and the k-th problem is a choice between two decisions whose gains are random variables G_k^1 and G_k^2, respectively. If $EG_k^1 - EG_k^2 \geq 0$ for every k, the difference $EG_k^1 - EG_k^2$ is reasonably large, n is not too small, and the variances of G_k^j's are not too large then $G_1^1 + \cdots + G_n^1$ is either truly or approximately stochastically larger than $G_1^2 + \cdots + G_n^2$. This conclusion is a mathematical theorem which requires precise assumptions, different from one case to another. Since $G_1^1 + \cdots + G_n^1$ is stochastically larger than $G_1^2 + \cdots + G_n^2$, we conclude that it is beneficial to maximize the expected gain in every of the n decision problems. This justification of the idea of maximizing of the expected gain does not refer to the Law of Large Numbers because it is not based on the approximate equality of $(G_1^j + \cdots + G_n^j)/n$ and its expectation. The number n of decision problems does not have to be large at all—the justification works for moderate n but the cutoff value for n depends significantly on the joint distribution of G_k^1's and G_k^2's.

(vi) An obvious drawback of the proposed decision-making philosophy is that not all decisions are comparable. Recall the utility function used by the subjective theory. I will reasonably assume that all utility functions are non-decreasing. It is easy to show that two decisions are comparable if and only if one of the decisions has greater expected utility than the other one for every non-decreasing utility function. Hence, the proposed ordering of decisions is consistent with the subjective philosophy in the following sense. In situations where the probabilities are undisputable, two decisions are comparable if and only if all decision-makers, with arbitrary non-decreasing utility functions, would make the same choice. Let me use the last remark as a pretext to point out a weakness in the subjective philosophy of probability. The comparability of all decisions in the subjective theory is an illusion because the ordering of decisions is strictly subjective, depending on the individual decision-maker. They can change the ordering of decisions by fiat at any time, so the ordering has hardly any meaning.

10.2.4 *Generating predictions*

My second proposal for a decision-making strategy is better adapted to laws (L1)-(L6), especially (L6), than the "stochastic ordering" presented in the previous subsection.

The basic idea is quite old—it goes back (at least) to Cournot in the first half of the nineteenth century (quoted after [15], p. 585):

> If the probability of an event is sufficiently small, one should act in a way as if this event will not occur at a solitary realization.

Cournot's recommendation contains no explicit message concerning events, which have probabilities different from 0 or 1. I propose limiting the probability-based decision making to the cases Cournot's assertion covers. I postulate that probabilistic and statistical analysis should make predictions its goal. In other words, I postulate that decision-makers should try to find events that are significant and have probabilities close to 1 or 0.

To illustrate the idea, I suppose that one faces a large number of independent decision problems, and at the k-th stage, one has a choice between decisions with payoffs G_k^1 and G_k^2, satisfying $EG_k^1 = x_1$, $EG_k^2 = x_2 < x_1$, $\mathrm{Var}G_k^j \leq 1$. If one chooses the first decision every time, the average gain for the first n

decisions will be approximately equal to x_1. The average will be approximately x_2 if one chooses the second decision every time. A consequence of the Large Deviations Principle is that the probability $P\big(\sum_{k=1}^{n} G_k^1/n \le (x_1+x_2)/2\big)$ goes to 0 exponentially fast as n goes to infinity, and so it can be assumed to be zero for all practical purposes, even for moderately large n. This and a similar estimate for $P\big(\sum_{k=1}^{n} G_k^2/n \ge (x_1+x_2)/2\big)$ generate the following prediction. Making the first decision n times will yield an average gain greater than $(x_1+x_2)/2$, and making the second decision n times will result in an average gain smaller than $(x_1+x_2)/2$, with probability p_n very close to 1. Here, "very close to 1" means that $1-p_n$ is exponentially small in n. The fast convergence rate is considered excellent in the present computer science-dominated intellectual climate.

The traditional curse of statistics is the slow rate (that is, $1/\sqrt{n}$) of convergence of approximations to the "true value," as indicated by the Central Limit Theorem. At the intuitive level, this means that to improve the accuracy of statistical analysis 10 times, one needs 100 times more data. The Large Deviations Principle, when used as in the above example yields a much better rate of convergence to the desirable goal. I conjecture that decision makers base their confidence in statistical methods because they subconsciously rely on predictions generated by the Large Deviations Principle rather than on claims derived from the Central Limit Theorem.

The decision-making strategy proposed in this section is partly based on the observation that in the course of real life, we routinely ignore events of minuscule probability, such as being hit by a falling meteor. Acting otherwise would make life unbearable and would be doomed to failure, as nobody could analyze all events of extremely small probability. Applying the Large Deviations Principle (consciously or subconsciously) can reduce the uncertainty to levels that are routinely ignored in normal life out of necessity.

The decision-making strategy proposed in this section yields applicable advice in fewer situations than the one proposed in the previous section. This strategy should be adopted by those who think that it is better to set goals for oneself that can be realistically and reliably attained rather than to deceive oneself into thinking that one can find a good recipe for success under any circumstances.

10.2.5 *Intermediate decision problems*

The "intermediate" in the title of this section refers to problems that do not fit into any of the following two templates.

Some events have probabilities so small that we assume that they will never occur, period. This is especially true if the possible losses and gains are moderate and so the expected value of gain or loss is negligibly small. A typical example of such an event is damage to one's property from a falling meteor. I do not think anyone takes precautions to protect their property against meteors. An almost universal decision is to ignore all very unlikely events.

Some decision situations can result in small gains or losses. A good example may be the choice of a restaurant for dinner. The gain or loss of utility due to a lucky or unlucky choice of a restaurant is usually considered small in comparison to other gains or losses in one's life. Individuals, groups of people, companies, etc. face very large numbers of such low-key decision problems in their lifetimes. Hence, one can apply the Law of Large Numbers in all such decision situations and hope that it gives a reasonably accurate prediction despite informal and approximate match between this mathematical theorem and reality. Maximization of the expected value of the gain in every case of decision problem at this level will result in the maximal gain over the lifetime of the decision maker (more precisely: decision beneficiary).

In both types of decision problems discussed above, the widely used strategies can be easily justified using (L6) because they both involve predictions, that is, events with very high probabilities.

"Intermediate" decision problems are those where we do not deal with isolated events of very small probability, and we do not deal with large numbers of decisions with small payoffs. Examples of intermediate decision problems include choices of college, profession, job, spouse and house made by an individual. A typical person makes only a handful of such choices in their lifetime. Each of such decisions may have a significant impact on one's wealth, success and happiness.

An intermediate decision problem cannot be reduced to finding a prediction (an event of very high probability) that would involve only the decision-maker. If many people use a decision strategy that maximizes the chances for lifelong happiness, the Law of Large Numbers implies that the percentage of happy people in the population will be maximized. For this to happen, all or at least many individuals have to identify their interests (or feelings) with those of a group of people. Such identification is known and common; it takes the form of loyalty. But loyalty is not universal, and I do not see any purely logical argument in favor of loyalty in decision making. Loyalty seems to be the result of evolution so this discussion is closely related to the topic of Sec. 10.3.

10.3 Decision making and resonance

It seems that animals do not make decisions in a sense relevant to our discussion. Animals have to make all kinds of decisions concerning food, shelter, safety, etc. But most animals seem to make all these decisions subconsciously. So far, researchers documented rather limited animal ability to consciously analyze information before making decisions—this applies to only a few species, mainly apes. We, humans, are animals and, just like other animals, we make many decisions in a purely instinctive manner.

Resonance is an ability that evolved through the Darwinian selection process. Our ability to make optimal subconscious decisions must have evolved through the same selection process because resonance that is not followed by good decisions is useless from the point of view of survival. This implies that our instinctive decision-making skills are fine-tuned, at least in the range of challenges that appeared during the process of evolution of human beings.

Our instinctive decision choices are often imperfect, just like the process of resonance is not perfect. Psychologists and economists collected a multitude of examples of suboptimal human behavior.

Conscious information processing geared towards decision making is needed for at least two major reasons. First, even "primitive" human societies were so complex that many decision problems were far outside the range that intuitive decision-making could handle well. The second reason is the constant pressure from individuals and groups of people who try to exploit instinctive decision making of other individuals for their own benefit, often to the detriment of the exploited individual.

I do not think that psychologists understand our decision-making instincts well. My pessimism is supported by a review of psychological research in [70]. So I will speculate that the evolution process shaped our minds so that we tend to make decisions that maximize the survival probability of the genealogical tree of descendants of the decision-maker. Or perhaps we subconsciously try to maximize the expected value of the number of descendants at a future time. I admit that both conjectures are speculative and heuristic. It is possible that the evolution process made us maximize some other functional of the universe. Assuming that the first of my conjectures is true, it would be natural to represent our subconscious decision-making preferences in terms of the utility function, which is simply the probability of survival—either survival of the individual or survival of the tree of descendants.

Even if all my speculations concerning the evolution of our subconscious decision making tendencies are correct, they do not provide a solution to the philosophical problem of choosing the best decision strategy. The fact that a specific decision strategy was favored by the evolution does not imply in any way that a rational person has to follow the same decision strategy. Consider the following analogy—if it is proved by psychologists and brain scientists beyond any doubt that evolution made us naturally aggressive, I doubt that anyone would consider this scientific finding a strong argument in favor of aggressive behavior.

Despite my apprehensive attitude towards adoption of maximization of the expected utility as the "rational" decision strategy, I realize that many people may be convinced of the value of this decision strategy by the fact that the evolution favored it and hence built it into our minds. However, this is no more than a choice. No logical argument can lead from the strategy favored by the evolution to our own free and conscious choice of the decision strategy.

10.4 Events with no probabilities

So far, I implicitly assumed that the relevant probabilities were known in my decision-making discussion. This section examines decision-making options in situations where (L1)-(L6) and available information do not determine the relevant probabilities.

One of the great and indisputable victories of the subjectivist propaganda machine is the widespread belief that there is always a way to choose a rational (or beneficial or the best) action in any situation involving uncertainty. Many of the people who otherwise do not agree with the subjective theory of probability, seem to think that it is a genuine intellectual achievement of the subjective theory to provide a framework for making decisions in the absence of relevant and useful information.

What can other sciences offer without information or relevant theories? A physicist cannot advise on how to build a plane flying at twice the speed of light or how to make a room temperature superconductor. Some things cannot be done because the laws of science prohibit them, and some things cannot be done because we have not learned how to do them yet (and perhaps we never will). Nobody expects a physicist to give "imperfect but adequate" advice in every situation—nobody knows how to build a plane that flies at "more or less" twice the speed of light or make a superconductor that works at "more or less" room temperature. No such leniency is shown towards probabilists and statisticians by people who take the subjectivist ideology seriously—if probability is subjective, then there is no situation in which you lack anything to make probability assignments. Moreover, you cannot be wrong if you are consistent.

What should one do in a situation involving uncertainty if no relevant information is available? An honest and rather obvious answer is that there are situations in which the probability theory has no scientific advice because no relevant probability laws or relations are known. This is not anything that we, probabilists, should be ashamed of.

The form of the laws (L1)-(L6) may illuminate the problem. The laws do not give a recipe for assigning values to all probabilities. They only say that the probabilities must satisfy some conditions in some circumstances. If no relevant relations, such as lack of physical influence or symmetry, are known, then laws (L1)-(L6) are not applicable, and any assignment of values to probabilities is arbitrary. Note that every event is involved in some relation listed in (L1)-(L6), for example, all events on the Earth are physically unrelated to a supernova explosion in a distant galaxy (except for some astronomical observations). Hence, strictly speaking, (L1)-(L6) are always applicable, but the point of the science of probability is to find sufficiently many relevant relations between events so that one can find good estimates for probabilities of relevant events.

One could argue that in a real-life situation, one must decide. Hence, one always (implicitly) assigns values to probabilities—in this limited sense, probability always exists. However, the same argument fails to establish that "useful relations between events can always be found." A practical situation may force a person to make a decision and, therefore, implicitly choose probability assignments. Still, nothing can force the person to make predictions that will eventually agree with observations. This reminds me of one of the known problems with torture (besides being inhumane): you can force every person to talk, but you do not know whether the person will be saying the truth.

On the practical side of the matter, it is clear that people use a lot of science in their everyday lives in an intuitive or instinctive way. Whenever we walk, lift objects, pour water, etc., we use laws of physics, more often than not at a subconscious level. We are pretty successful with these informal applications of science, although not always so. The same applies to probability—a combination of intuition, instinct, and reasoning based on analogy and continuity can give excellent practical results. This, however, cannot be taken as proof that one can always assign values to all probabilities and attack every decision problem in a scientifically justified way.

Chapter 11

Frequency Statistics

It is a common view that frequency statistics is justified by the "frequency philosophy." Writing a chapter proving that frequency statistics is unrelated to the philosophical theory of von Mises is pointless. His theory of collectives was abandoned a long time ago. I guess that less than one percent of frequency statisticians know what a collective is. However, I cannot summarily dismiss the idea that the "frequency interpretation" of probability is the basis of frequency statistics. There is a difficulty, though, with analyzing the "frequency interpretation"—unlike von Mises' theory, the frequency interpretation is a mixture of mathematical theorems and intuitive feelings, not a fully developed philosophy of probability. Despite this problem, I will try to give a fair account of the relationship between the frequency statistics and the frequency interpretation of probability.

Three popular methods developed by frequency statisticians are estimation, hypothesis testing and confidence intervals.

11.1 Confidence intervals

The concepts of an estimator and a hypothesis test seem to be more fundamental to frequency statistics than the concept of a confidence interval. First, I will discuss confidence intervals because I will refer to some of this material in the section on estimation.

Suppose that a parameter θ, presumably an objective

physical quantity, is unknown, but some data related to this quantity are available. A "95%-confidence interval" is an interval constructed by a frequency statistician based on the data, and such that the actual value of the parameter θ is covered by the interval with probability 95%. More precisely, statisticians prove that if the value of the unknown parameter θ is θ_0 then the 95% confidence interval will contain θ_0 with probability 95%.

Among the basic methods of frequency statistics, confidence intervals fit (L1)-(L6) best because they are probability statements, unlike estimators and hypothesis tests. Hence, a single confidence interval is a prediction in the sense of (L6), if the probability of coverage of the unknown parameter is chosen to be very high.

I would call a confidence interval a prediction only if the probability of the coverage of the true value of the parameter was 99% or higher. This does not mean that we cannot generate predictions when we use confidence intervals with lower probability of coverage—we can aggregate multiple cases of confidence intervals and generate a single prediction. A scientist or a company might not be interested in the performance of a confidence interval in a single statistical problem, but the performance of an aggregate of statistical problems.

Suppose that n independent 95% confidence interval are constructed. If n is sufficiently large, then one can make a prediction that at least 94% of the intervals will cover the true values of the parameters, with probability 99.9% or higher. No matter where we draw the line for the confidence level for a single prediction, we can generate a prediction at an arbitrarily high confidence level by aggregating a sufficiently large number of confidence intervals.

Another way to analyze the scientific performance of confidence intervals is to express losses due to non-coverage errors using the units of money or utility. We can apply, at least in principle, one of the probabilistic techniques to find the distribution of the aggregate loss due to non-coverage errors for a family of confidence intervals and generate a corresponding prediction, for example, a single 99.9% confidence interval for the combined loss.

11.2 Estimation

The theory of estimation is concerned with unknown quantities called parameters. Examples of such quantities include the speed of light, the probability of a side effect for a given drug, and the volatility in a financial market.

Let us consider a simple example. If you toss a deformed coin, the results may be represented by an i.i.d. (independent identically distributed) sequence of heads and tails. The probability of heads (on a single toss) is an unknown constant (parameter) θ, and the statistical analysis aims to find a good estimate of the true value of θ. If n tosses were performed, and k of them resulted in heads, then one can take k/n as an estimate of θ. This estimator (that is, a function generating an estimate from the data) is unbiased in the sense that the expected value of the estimate is the true value of θ.

The ultimate theoretical goal of the estimation theory is to find an explicit formula for the distribution of a given estimator for each parameter value θ. This, theoretically, allows one to derive all other properties of the estimator because the distribution encapsulates all the information about the estimator. Quite often, an explicit or even approximate formula for the distribution is impossible to derive. In such cases, one can try to prove a weaker property of the estimator, for example, that it is unbiased, which means that its expected value is equal to the true value of the parameter.

Suppose that an estimator is unbiased. The long-run interpretation of this statement requires that we collect a long sequence of data sets, all in the same manner, and independently from each other. A crucial assumption is that the parameter (presumably, a physical quantity) is known to have the same value every time we collect a set of data. Moreover, suppose we apply the same estimator every time we collect a data set.

Then, according to the Law of Large Numbers, the average of the estimates will be close to the actual value of the unknown parameter with a high probability. The scenario described above is purely imaginary. There are many practical situations when one of the conditions described above holds. For example, multiple estimates of the same quantity are sometimes made (think about estimating the speed of light). There are also applications of the estimation theory when the same estimator is applied routinely over long periods; for example, a medical laboratory may estimate the hormone level for a large number of patients. It hardly ever happens that data sets are repeatedly collected in an identical way, and a long sequence of isomorphic estimates are calculated, assuming that the estimated quantity is known to have the same value in each case. Social scientists often repeatedly measure the same quantity characterizing people or society, but their measurement methods and definitions often differ. The number of repetitions is hardly ever very large.

The long-run frequency interpretation of unbiasedness is unrealistic for another somewhat different reason. Statisticians often work with only one data set at a time, and there is no explicit or implicit expectation that isomorphic data sets will be collected in the future. I do not think that any statistician would feel that the value of a given (single) estimate diminished if they learned from a clairvoyant person that no more similar estimates would ever be made. It will not help to imagine a long sequence of similar estimation problems.

I have already explained why the frequency interpretation does not explain why a specific property of estimators, unbiasedness, is useful in practice. I will now give several reasons why the frequency interpretation does not support the statistical theory of estimation in general.

When multiple data sets, measurements of the same physical quantity are collected, they often differ by their size (number of data)—this alone disqualifies a sequence of data sets from being an i.i.d. sequence, or a collective in the sense of von Mises.

Frequency statisticians do not require that estimators are

applied only to very large data sets. The accuracy of the estimator depends on the amount of data, and, of course, the larger the number of data, the better the accuracy. This is never taken to mean that the estimator cannot be used for a small number of data. It is left to the end-user of statistical methods to decide whether the estimator is useful given any particular amount of data. This indicates that the idea of von Mises that only very long sequences should be considered is often ignored.

Frequency statisticians do not hesitate to analyze models which are far from von Mises' collectives, for example, stationary processes and Markov processes. Some stochastic processes involve complex dependence between their values at different times, so they are far from collectives. Some families of stochastic processes can be parameterized; for example, distinct members of a family of stationary processes may be labeled by one or several real numbers. Nothing would prevent a frequency statistician from estimating parameters of a process in this family. A frequency statistician would feel no obligation to find an i.i.d. sequence of trajectories of the process. A single trajectory of a stochastic process can be the basis for estimation in frequency statistics.

11.2.1 *Estimation and (L1)-(L6)*

The theory of estimation can be justified in several ways using (L1)-(L6), just like the theory of confidence intervals.

Consider a single statistical problem of estimation. Recall that, according to (L6), a verifiable statistical statement is a prediction, that is, an event of probability close to 1. The only practical way to generate a prediction from an estimator is to construct a confidence interval using this random variable. When we estimate several parameters at the same time or in the infinite-dimensional ("non-parametric") case, a confidence interval has to be replaced with a "confidence set," that is, a subset of a large abstract space.

Another justification based on (L1)-(L6) for the use of

estimators comes from aggregate statistical problems. Suppose that n independent estimates are made, and each is used to construct a 95% confidence interval. If n is sufficiently large, then one can make a prediction that at least 94% of the intervals will cover the true values of the parameters, with probability 99.9% or higher.

Finally, just like in case of confidence intervals, we may apply the decision-theoretic approach. We may express losses due to errors using the units of money or utility. Once the losses are expressed in the units of money or utility, one can apply, at least in principle, one of the probabilistic techniques to find the distribution of the aggregate loss and generate a corresponding prediction, for example, a 99.9% confidence interval for the combined loss.

11.3 Hypothesis testing

The estimation theory, a part of frequency statistics, has nothing in common with von Mises' theory of collectives. It cannot be justified by the "frequency interpretation" of probability, except in very few exceptional cases. The relationship between hypothesis testing (see Sec. 3.2) and von Mises' collectives, and the relationship between hypothesis testing and the frequency interpretation of probability are more subtle.

Hypothesis tests are prone to two kinds of errors. One can make a Type I error also known as "false positive", that is, the incorrect rejection of the null hypothesis. The other possible error is called Type II error or "false negative", that is, the incorrect acceptance of the null hypothesis.

There are at least three common settings for hypothesis testing. First, in some industrial testing applications, one has to test many identical items for defects. The probabilities of false negatives and false positives have a clear interpretation as long-run frequencies in such situations. Another good example of a long run of isomorphic hypothesis tests is a series of medical tests, say, for HIV.

In a scientific laboratory, a different situation may arise. A scientist may want to find a chemical substance with desirable properties, say, a new drug. They may perform multiple hypothesis tests on various substances and the models involved in these testing problems may differ greatly. In this case, a statistical prediction would involve the percentage of a hypothesis's false rejections or false acceptances. Just like in the case of estimation, one cannot use the von Mises theory of collectives in this case because the tests are not necessarily identical. Nevertheless, if we assume that they are independent then we can make a prediction about the aggregate rate of false rejections or aggregate rate of false acceptances.

The third type of situation when hypothesis tests are used is when a single hypothesis is tested, with no intention to relate this test to any other hypothesis test. A good example is a criminal trial. In the US, a defendant's guilt has to be proved "beyond a reasonable doubt." Since the jury has only two choices—guilty or not guilty—criminal trials are examples of hypothesis testing, even if rarely (likely never) are they formalized using the statistical theory of hypothesis testing. If a jury votes "guilty," it effectively makes a prediction that the defendant committed the crime. In this case, the prediction refers to an event in the past. In a criminal trial, a prediction is the result of a single case of hypothesis testing. A long-run frequency interpretation of hypothesis testing in case of criminal trials is possible but has a questionable ethical status. Presumably, in a given criminal trial, the society expects the jury to make every effort to arrive at the right decision in this specific case.

11.3.1 *Hypothesis tests and the frequency interpretation of probability*

The frequency interpretation of probability does apply to hypothesis tests and, in theory, provides a good philosophical support for this method.

Recall the two situations where long runs of hypothesis tests are applied. First, we may have a long run of isomorphic tests, with "simple" null and alternative hypotheses. For example, a machine part can be declared either defective or not. Then, one can predict the frequency of false positives, that is, false classifications of parts as defective. Similarly, we can predict the frequency of false negatives. Both frequencies can be observed in practice in some situations, and this can be easily turned into a prediction in the sense of (L6). Hence, the theory of hypothesis testing is a well-justified science.

A subtle point here is that both frequencies, of false positives, and false negatives, are within sequences unknown to the statistician. This is because the statistician does not know in which tests the null hypothesis is true. Hence, the predicted frequencies are not directly observable in many situations. One could argue that the frequencies of false positives and false negatives may be observable in the future when improved technology allows us to revisit old tests and determine which null hypotheses were true. Strictly speaking, this turns the predicted frequencies of false positives and false negatives into scientifically verifiable statements. However, this does not necessarily imply that predictions verifiable in this sense are helpful in practice. Hence, it is rational to recognize the theory of hypothesis testing as a scientific theory (in the sense given above) but to reject it in some practical situations because its predictions are not helpful.

11.3.2 *Hypothesis testing and (L1)-(L6)*

Generally speaking, hypothesis testing fits into the framework of (L1)-(L6) just like confidence intervals and estimators do. One can try to generate predictions in the sense of (L6) based on a single hypothesis test, or a sequence of hypothesis tests. There are some differences, though.

Sequences of hypothesis tests

Consider a sequence of hypothesis tests and, for simplicity, assume that the significance level is 5% in each case. Moreover, suppose the tests are independent but not necessarily isomorphic. Then, in the long run, the percentage of false positives, that is, cases when the null hypothesis is true, but it is rejected, will be about 5%. This is (or rather can be transformed into) a scientific prediction, in the sense of (L6). Let me repeat some remarks from the last section to ensure my claim is not misunderstood. Statisticians do not know which null hypotheses are true when they perform the tests. For example, they may learn this later when better technology is available. In case of a medical test for a virus, the condition of a patient may change soon so that it will be known whether she has the virus or not. The time in the future, when we learn whether the null hypothesis is true or not with certainty, may be very distant, depending on the specific problem. Hence, the prediction, although scientifically verifiable in the abstract sense, may be considered irrelevant because of the time delay. A prediction concerning false negatives can be generated similarly, but this is a bit more complicated—I will try to outline some difficulties below.

There are some practical situations when the null hypothesis and the alternative hypothesis are simple; for example, a manufacturer may want to classify parts as defective or non-defective. However, there are also practical situations where hypotheses are not simple, for example, when the rates of side effects for two drugs are either identical (the null hypothesis) or they differ by a real number between -1 and 1 (the alternative). The issue of generating a prediction for false negatives in the long run setting is complicated by the fact that it is not obvious how to choose the appropriate sequence of tests in which a predicted frequency of false negatives will be observed. Should we consider all cases when the two rates of side effects are different? In theory, this would mean all tests, as it is virtually impossible that the difference between the rates is exactly zero. Or should we consider only those cases when the difference of rates is greater than some fixed number, say, 5%?

Single hypothesis test

A single hypothesis test can be considered an application of (L6) in its pure form. We incorporate the null hypothesis into our model. Then we make a prediction that a certain random variable (the "test statistic") will not take a value in a specific range. We observe the value of the test statistic. If the value is in the "rejection region," we conclude that something was wrong with the theory that generated the prediction, which usually means that we reject the null hypothesis.

The above algorithm correctly applies (L6), and shows that a single hypothesis test is a scientific procedure. There remain some issues that need to be addressed, though. The first is the question of the significance level. A popular significance level is 5%. In other words, a hypothesis test implicitly assumes that an event is a prediction if its probability is 95% or higher. The choice of probability which makes an event a prediction is subjective. Personally, I find 95% too low. I would hesitate to draw strong conclusions if an event of probability 95% failed. I would prefer to raise the level to 99%, but this choice makes effective hypothesis testing more costly.

Second, the interpretation of hypothesis testing described above, as a pure application of (L6), may be too crude for scientific purposes. Consider applications of geometry. We can consider an automobile wheel a circle for some practical purposes. For other purposes, engineers have to describe its shape with much greater precision. They would conclude that a wheel is not a circle when accurately measured. In the simplest setting, the result of testing a hypothesis is its rejection or acceptance. Hence, hypothesis testing makes an impression of a crude method. In practice, it is often supplemented by the "p-value," an extra piece of information for the statistician and end users of the test. The p-value is the probability that we observe the actual data or "more extreme" data if the null hypothesis is true. The p-value does not have a simple frequency interpretation, although frequentists use it. It is very unlikely that a sequence of isomorphic hypothesis tests will yield identical or very similar p-values. To put it differently, it would be very unusual for a collection of hypothesis tests grouped together because they yielded very similar p-values to have a practical application.

Chapter 12

Bayesian Statistics

Bayesian statistics is a very successful branch of science because it is capable of making excellent predictions, in the sense of (L6).

The general structure of Bayesian analysis is universal—the same scheme applies to all cases of statistical analysis. One of the elements of the initial setup is a "prior," that is, a prior probability distribution, a consistent view of the world. The second element is the data from an experiment or observations. A Bayesian statistician then applies the Bayes theorem to derive the "posterior," that is the posterior probability distribution, a new consistent view of the world. The posterior can be used to make decisions—one has to find the expected value of the gain associated with every possible decision and take the action that maximizes this expectation.

12.1 Elements of Bayesian analysis

The simple and clear structure of Bayesian statistics outlined at the beginning of the chapter conceals an important difference between philosophical "priors" and Bayesian "priors." In the subjective philosophy, the prior is a complete probabilistic description of the universe, used before the data are collected. In Bayesian statistics, the same complete description of the universe is split into a "model" and a "prior" (see example below). These poor linguistic practices lead to considerable confusion. Some people believe that using (statistical) subjective priors is a legitimate scientific practice, but the same people would not be willing to accept the use of subjective models. On the philosophical side, there is no demarcation line between subjective priors and subjective models.

Consider tosses of a deformed coin—this simple example will illustrate the statistical usage of the words "prior" and "model." The prior distribution will be specified in two steps. First, a "model" will be found. The model will involve some unknown numbers, called "parameters." The term "prior" refers in Bayesian statistics only to the unknown distribution of the "parameters." In a sequence of coin tosses, the results are usually represented mathematically as an exchangeable sequence. According to de Finetti's theorem, an exchangeable sequence is equivalent (mathematically) to a mixture of "i.i.d." sequences. Here, an "i.i.d. sequence" refers to a sequence of independent tosses of a coin with a fixed probability of heads. The assumption that the sequence of tosses is exchangeable is a "model." This model does not uniquely specify which i.i.d. sequences enter the mixture and with what weights. The mixing distribution (that is, the information on which i.i.d. sequences are a part of the mixture, and with what weights), and only this distribution, is customarily referred to as the "prior" in Bayesian statistics.

12.2 Models

12.2.1 *Bayesian models are totally objective*

In Bayesian analysis, models are treated as objective representations of objective reality. One of the common misunderstandings about the meaning of the word "subjective" comes here into play. Bayesian statisticians may differ in their opinions about a particular model that would fit a particular real-life situation—in this sense, their views are subjective. For example, some of them may think that the distribution of a given random variable is symmetric, and others may have the opposite opinion. This kind of subjectivity has nothing to do with de Finetti's subjectivity—according to his theory, symmetry in the real world, even if it is objective, is not linked in any way to probabilities because probability values cannot be objectively determined. Hence, according to the subjective theory, differences in views between Bayesian statisticians on a particular model are irrelevant from the point of view of the future success of the statistical analysis—no matter what happens, nothing will prove that any particular model is right or wrong. I do not find even a hint of this attitude among Bayesian statisticians. The importance of matching the model to the real world is taken as seriously in Bayesian statistics as in frequency statistics. Bayesian statisticians think that it is a good idea to make their mathematical model symmetric if the corresponding natural phenomenon is symmetric. In other words, they act as if they believed in objective probability relations. Bayesian models are based on (L1)-(L6) and other laws, specific to each case of statistical analysis.

12.2.2 *Bayesian models are totally subjective*

See Sec. 9.1.6 for the initial part of this discussion. Bayesian models are subjective because the prior can be corrected by the data but the model typically is not corrected by the data. The model arises from resonance, and we can only have faith that it is correct. Of course, models can be tested and are tested in statistical practice in some cases, but then untested resonance is a component of the verification procedure for the given model. In this specific and narrow sense, Bayesians developed inverted beliefs about what (prior or model) is objective and subjective in their analysis.

12.3 Priors

One could expect that of all the elements of the Bayesian method, the prior distribution would be the most subjective. Recall that in practice, the term "prior distribution" refers only to the opinion about the "unknown parameters," that is, that part of the model, which is not determined by (L1)-(L6) or some other considerations specific to the problem in a way that can be considered objective.

There are strong indications that priors are not considered subjective and that they do not play the role assigned to them by the subjective theory. One of them is a common practice to chose the prior after collecting the data. This disagrees with the subjective ideology in several ways. From the practical point of view, one can suspect that the prior is tailored to achieve a particular result. From the subjectivist point of view, the prior is meant to represent decisions made before collecting the data—the fact that the prior is often chosen after collecting the data shows that there were no relevant decisions made before collecting the data and so there is no need to coordinate anything.

Surprisingly, Bayesian statisticians discuss the merits of different prior distributions. This suggests they do not believe in the subjectivity of priors. If the prior is subjective in the personal sense, if it reflects one's own opinion, then there is nothing to discuss—the prior is what it is. Moreover, deliberations of various properties of priors indicate that some priors may have some demonstrably good properties—this contradicts the spirit and the letter of the subjective theory.

According to the subjective theory, no subjective prior can be shown to be more true than any other prior. Hence, one could benefit by simplifying the prior. Many priors can save money and time by reducing the computational complexity of a problem. For example, we could assume that the future events are independent from the data and the past. Then the Bayes theorem implies that the posterior distribution is equal to the prior distribution. Hence, one does not have to do any calculations—the savings of time and money can be enormous. In the context of deformed coin tossing, a very convenient prior is the one that makes the sequence of coin tosses i.i.d. with the probability of heads equal to 70%. This subjective opinion does not require Bayesian updating when the results of coin tossing are observed—the posterior is always the same as the prior. Needless to say, in Bayesian statistics, priors are never chosen just based on their technical complexity. Bayesian statisticians believe that they benefit objectively by rejecting simplistic but computationally convenient priors.

Bayesian statisticians choose priors to obtain the most reliable predictions based on the posterior distributions. The matter is somewhat complicated by mathematical and technical limitations. Not all priors lead to tractable mathematical formulas, and some priors require an enormous amount of computer time to be processed. Setting these considerations aside, we can distinguish at least three popular ways of choosing priors in Bayesian statistics: (i) an application of (L1)-(L6), (ii) a technically convenient probability distribution, not pretending to represent any real probabilities, and (iii) informal summary of statistician's knowledge. I will discuss these choices in more detail.

12.3.1 *Objective priors*

The adjective "objective" in the title of this subsection indicates that some priors involve probability assignments that can be objectively verified, for example, by long run repetitions that do not involve any Bayes-related reasoning.

Textbook examples show how one can choose a prior using (L1)-(L6). Suppose that there are two urns; the first contains 2 black and 7 white balls, and the second one contains 5 black and 4 white balls. Someone tosses a coin and samples a ball from the first urn if the result of the toss is heads. Otherwise, a ball is sampled from the second urn. Suppose that the color of the sampled ball is black. What is the probability that the coin toss resulted in tails? In this problem, the prior distribution assigns equal probabilities to tails and heads by symmetry. More interesting situations arise when long-run frequencies are available. For example, suppose that 1% of the population in a certain country is infected with HIV, and an HIV test generates

1% false negatives and 10% false positives. (A false positive is when someone does not have HIV, but the test says he does.) If someone tests positive, what is the probability that he has HIV? In this case, the prior distribution says that a person has HIV with probability 1%. This is based on the long run frequency and, therefore, implicitly on (L1)-(L6). In other words, a symmetry is applied because the tested person is indistinguishable, to the tester, from other people in the population.

12.3.2 *Bayesian statistics as iterative method*

Some priors play the role of the "seed" in an iterative method. Such methods are popular in mathematics and numerical analysis.

Suppose we want to find a solution to a differential equation, that is, a function that solves the equation. An iterative method starts with a function S_1, that is, a "seed." Then one has to specify an appropriate transformation $S_1 \to S_2$ that takes S_1 into a function S_2. Usually, the same transformation is used to map S_2 onto S_3, and so on. The method works well if one can prove that the sequence $S_1, S_2, S_3, \ldots$ converges to the desirable limit, that is, the solution of the differential equation. The convergence has been proved in many cases. This iterative method is used in both pure mathematics and applied numerical methods. The seed S_1 is not assumed or expected to be a solution to the differential equation or even close to such a solution. Not all seeds will generate a sequence of S_k's converging fast to the desirable limit, and the number of iterations needed for a good approximation of the solution depends on the problem itself and the seed. The choice of an efficient transformation $S_k \to S_{k+1}$ and the seed S_1 is a non-trivial problem with no general solution—the answer depends on the specific situation.

The Bayesian method can be interpreted as an iterative scheme when the number of data is reasonably large. The following two procedures are equivalent from the mathematical point of view. The standard algorithm is to combine the prior distribution with all of the available data, using the Bayes theorem, to obtain the posterior distribution. An alternative, mathematically equivalent representation is to start by combining the prior distribution with a single piece of data to obtain an intermediate distribution. This new distribution can be combined with another single piece of data to obtain another intermediate distribution, and so on. When we finish the process by including the last piece of the data, the resulting distribution will be the same as the one obtained in one swoop.

The general success of iterative methods suggests that Bayesian statistics might be successful because it can be represented as an iterative scheme. This is indeed the case in typical situations. I will illustrate the claim with a simple example. Consider tosses of a deformed coin. It is popular to use the "uniform" prior, that is, to assume that the sequence of tosses is a mixture of i.i.d. sequences, each i.i.d. sequence represents a coin with the probability of heads equal to p, and p itself is a random variable which lies in any subinterval $[a, b]$ of $[0, 1]$ with probability $b - a$. If there were k heads in the first n trials, then the probability of heads on the $(n+1)$-st toss is $(k+1)/(n+2)$. For large n, this is very close to k/n, a value that many people would consider a good intuitive estimate of the probability of heads on the $(n+1)$-st toss. The reliability of the estimate $(k+1)/(n+2)$ can be confirmed by real data.

The uniform prior in the last example does not represent any "subjective opinion" and it does not represent any "objective probability" either. It is a seed in an iterative method, and it works in an almost magical way—no matter what your coin is, you can expect the posterior distribution to yield excellent probability estimates. I said "magical" because, in mathematics in general, there is no reason to expect every iterative method and seed to be equally efficient. The practice of using the prior distribution as an abstract seed in an iterative method is perfectly well justified by (L1)-(L6), because one can empirically verify predictions implicit in the posterior distribution, in the spirit of (L6).

An important lesson from the representation of some Bayesian algorithms as an iterative method is that they may yield little useful information if the data set is not large. What this means depends, of course, on the specific situation. Most people assume, at least implicitly, that the value of the posterior is almost negligible when the model and the prior are not based on (L1)-(L6) or the data set is small. The subjective philosophy makes no distinction whatsoever between probability values arrived at in various ways—they are all equally subjective and unverifiable.

12.3.3 *Truly subjective priors*

A prior may represent a scientist's prior knowledge informally. Using such priors is a sound scientific practice, but this practice might be the most misunderstood element of Bayesian statistics. Statisticians know that using personal prior distributions often yields excellent results. Hence, a common view is that subjective priors "work," and hence the subjective theory of probability is vindicated.

It is clear that Bayesian statisticians expect posterior distributions to generate reliable predictions. Hence, the question is why and how personal priors can be the basis of objectively verifiable predictions? Resonance is the source and the explanation of reliable personal priors. In the Bayesian context, resonance is needed to construct reliable models and it is also needed to generate reasonable prior distributions. I will discuss two scenarios of statistical analysis—one with small amount of data and the other one with a large amount of data. Before discussing these scenarios, let me recall an elementary fact from Bayesian statistics—in a typical statistical model, no matter what data are, a suitable prior distribution can totally overwhelm the data and almost completely determine the posterior distribution. Hence, the choice of the prior distribution cannot be arbitrary—that would lead to nonsensical statistical inference.

If the amount of data is small, the prior distribution determines the posterior distribution to a large extent and hence must be chosen carefully. The prior opinion of a scientist may be based on pure resonance. The prior distribution may also be the result of informal processing of information that, under the best of circumstances, could have been processed explicitly and formally. The prior distribution generated in this way may be somewhat different from the distribution that would have been generated from the same information by formal mathematical calculations. Nevertheless, that does not mean that such an informal prior is completely different from the results of a formal calculation. This is similar to the informal assessment of temperature. We cannot precisely measure the outside temperature using our senses, but that does not mean that our subjective opinions about temperature are hopelessly inaccurate or useless. Hence, some personal priors work because they are not much different from "objective priors" discussed in Sec. 12.3.1. Our resonance abilities, at different levels and acquired in different ways, explain our success in choosing priors that lead to reliable posterior distributions and predictions based on them.

Another reason why priors representing personal knowledge often work well is that sometimes they are combined with large amounts of data. In such a case, the posterior distribution is not very sensitive to the choice of the prior. In other words, in such cases, the personal prior plays the role of the seed in an iterative method, and quite often, its intrinsic value is irrelevant. This does not mean that the personal prior can be arbitrary. Recall that every data set can be overwhelmed by some prior distribution. Hence, resonance is needed to specify a prior that will let the data determine the posterior distribution. If the amount of data is large, the influence of the prior on the posterior is small. This means resonance is needed to eliminate only the most extreme priors. The collection of reasonable priors is large, and resonance is needed only to find a prior in this large collection.

Priors that summarize personal knowledge informally are not guaranteed to generate reliable posterior distributions. They summarize what cannot be built into the model—this alone is suspicious. They are prone to prejudices of various kinds, from scientific to social and personal.

The subjective ideology is harmful to Bayesian statistics because it muddles the distinction between priors, which represent information gathered and processed in an informal way and priors that are objective or are seeds of an iterative algorithm. It has been proposed that some priors, for example, uniform priors, represent the "lack of knowledge." This may be an intuitively appealing idea, but it is hard to see why a prior representing the "lack of knowledge" is useful in any way. The sole test of the value of a prior is the quality of predictions generated by the posterior distribution corresponding to the prior.

Consider the following example to see the actual role of the prior. Suppose that a scientist has to make a decision. She uses the Bayesian approach, chooses a prior, and collects some data. When she is finished, the computer memory fails, she loses all the data, and she has no time to collect any more data before making a decision. According to the subjective philosophy, the prior represents the basis for the best course of action in the absence of any additional information. Hence, according to the subjective theory, the statistician has to make a decision based on her prior. In reality, nobody seems to be trying to choose a prior considering the potential disaster described above. If anything like this ever happened, there would be no expectation that the prior chosen to fit with the whole statistical process (including data collection and analysis) would be useful in any way without data.

12.4 Resonance at work

Bayesian statistics suffers from the confusion caused by the lack of understanding of resonance but provides an excellent opportunity to illustrate how resonance works in practice in a non-trivial situation.

Statistical and, in general, scientific theories are often

presented as models with parameters. The model is obtained through the process of resonance. Once the model is constructed, it can be analyzed by the usual scientific process of falsification/verification based on logic. The space of models is discrete in the following sense. If a model is rejected, there is no reason to believe that a better model for the same phenomenon should be "close" to the original one, although it could.

The parameters turn each model into a vector space or manifold (a continuous structure). The process of resonance is not sufficiently precise to provide the values of the parameters. But resonance gives bounds for the values of the parameters. The Bayesian prior is a way to encode beliefs based on resonance about the part of the parameter space where the parameter values reside. Sometimes, no useful bounds can be found, and the support of the prior distribution is the whole parameter space, as in the case of the uniform distribution. Some priors, for example Jeffrey's priors try to encode some general beliefs obtained through resonance, for example, beliefs about scaling and symmetry.

Models are discrete, and parameters are continuous in the following sense. Models are supposed to be accepted or rejected. Acceptance does not mean perfection, just sufficient applicability. Parameters are supposed to be adjusted. The best values are chosen among infinitely many values in the space.

A model is always based on resonance. Sometimes the model can be based on raw data with no understanding of the laws of nature underlying the phenomenon. This was the case of diffusion models for stock market prices in the past. More recently, similar models were derived from basic assumptions about the behavior of investors.

12.5 Data

It sometimes happens that the observed data do not seem to fit the model at all. In its pure subjectivist version, the Bayesian approach is totally inflexible—the posterior distribution must be derived from the prior and the data using the Bayes theorem, no matter what the prior and the data are. In practice, when the data do not match the model, the model is modified. This practice is well justified by (L1)-(L6) as an attempt to improve the reliability of predictions. The subjective philosophy does not justify changing the model or the prior once the data are collected.

Paradoxically, the subjective philosophy provides no support for the idea that it is better to have a lot of data than to have little data—see Sec. 6.11. Since the subjective philosophy is only concerned with the consistency of decisions, collecting data is not needed. Collecting data will not make it any easier for the decision-maker to be consistent than in the case when he has little or no data. Needless to say, Bayesian statisticians believe that collecting data is beneficial.

12.6 Posteriors

The posterior has the least subjective status of all elements of Bayesian statistics, mainly because of the social pressure. Business people, scientists, and ordinary people would have nothing to do with a theory that emphasized the subjective nature of its advice. Hence, the subjectivity of (some) priors is mentioned occasionally, but posterior distributions are implicitly advertised as objective. Take, for example, the title of a classical textbook [16] on Bayesian statistics, "Optimal Statistical Decisions." Optimal? According to the subjective theory of probability, your opinions can be either consistent or inconsistent—they cannot be true or false, and hence your decisions cannot be optimal or suboptimal. Of course, you may consider your own decisions optimal, but this does not say anything beyond the fact that you have not found any inconsistencies in your views—the optimality of your decisions is tautological. Decisions may also be optimal in some purely mathematical sense, but I doubt that was the intention of DeGroot when he chose the title for his book. The title was consciously or subconsciously chosen to suggest objective optimality of Bayesian decisions.

The posterior distribution is the result of combining the prior and the model with the data. Quite often, the prior is not objective (see Sec. 12.3.1) so the posterior is not based on (L1)-(L6) alone. This is one reason why posterior probability assignments are not always correct, in the sense of predictions, as in (L6). The weakest point of the philosophical foundations of Bayesian statistics is that they do not stress the necessity of a proof (in the sense of (L6)) that the posterior distribution has desirable properties. The subjective philosophy not only fails to make such a recommendation but asserts that this cannot be done at all. Needless to say, Bayesian statisticians routinely ignore this part of the subjective philosophy and verify the validity of their models, priors and posteriors in various ways.

From time to time, somebody expresses an opinion that the successes of Bayesian statistics vindicate the claims of the subjective philosophy. The irony is that according to the subjective theory itself, nothing can confirm any probabilistic claims—the only successes that the Bayesians could claim are consistency and elimination of Dutch books—this alone would hardly make much of an impression on anyone.

12.6.1 *Non-convergence of posterior distributions*

One of the most profound misconceptions about de Finetti's theory and Bayesian statistics is the claim that even if two people have two different prior distributions then their posterior distributions will be closer and closer to each other as the number of available data grows larger and larger. This misunderstanding is firmly rooted in the ambiguity of the word "prior." The philosophical "prior" includes the statistical "prior" and "model."

The primary example of a situation when Bayesian posterior distributions of two people will converge when more and more data are available is when the data sequence is objectively exchangeable. This claim is supported by empirical data. Some people believe this argument proves the value of subjectivism.

Consider the following example. Let X_k be the number of heads minus the number of tails in the first k tosses of a coin. Let Y_k be 0 if X_k is less than 0 and let Y_k be equal to 1 otherwise. The sequence $X_1, X_2, X_3, \ldots$ is known as a simple random walk, and the sequence $Y_1, Y_2, Y_3, \ldots$ is known not to be exchangeable. Suppose two people are shown some values of $Y_1, Y_2, Y_3, \ldots$, but only one knows how this sequence of zeroes and ones was generated. The other person might assume that $Y_1, Y_2, Y_3, \ldots$ is exchangeable, not knowing anything about its origin. If he does so, the posterior distributions of the two observers will not converge to each other as the number of observations grows.

The above elementary example is quite typical. If two statisticians disagree on the model then there is no reason to think that their posterior distributions will be close to each other, no matter how much data they collect. Conversely, if two statisticians adopted the same model but used two different priors then their posterior distributions will be closer and closer as the number of data grows, under mild technical assumptions.

12.7 Bayesian statistics and (L1)-(L6)

Methods of Bayesian statistics can be justified just like the methods of frequency statistics—see Sec. 11.1 and Sec. 11.2.1. Briefly, Bayesian statisticians can generate predictions in the sense of (L6). These can take the form of confidence intervals (called "credible intervals" in the Bayesian context). Predictions can be also based on aggregates of statistical problems, or can be made using decision theoretic approach.

Bayesian statisticians may use their own methods of evaluating their techniques as long as users of Bayesian statistics are satisfied. Predictions, in the sense of (L6), are needed because of the existing controversy within the field of statistics. Critics of Bayesian statistics must be given a chance to falsify predictions made by Bayesian statisticians. If the critics fail to falsify them, then, and only then, Bayesian statisticians may claim that their methods form a solid branch of science.

12.8 Who needs subjectivism?

There are (at least) two reasons why some Bayesian statisticians embrace the subjective philosophy. One is the mistaken belief that, in some cases, there is no scientific justification for the use of the prior distribution except that it represents the subjective views of the decision maker. I argued that the prior is sometimes based on (L1)-(L6) and other forms of resonance, sometimes it is the seed of an iterative method, and sometimes it is the result of informal processing of information. The justification for all of these choices of the prior is quite simple—predictions based on posterior distributions seem to be reliable.

Another reason for the popularity of the subjective philosophy among some Bayesians is that the subjective theory provides an excellent excuse for using the expected value of the (utility of) gain as the only determinant of the value of a decision. This is an illusion based on a clever linguistic manipulation—the identification of decisions and probabilities is true only by a philosopher's fiat. If probabilities are derived from decisions, there is no reason to think that they represent anything in the real world. The argument in support of using the expected value is circular—probabilities and expectations are used to encode a rational choice of decisions and then decisions are justified by appealing to thus generated probabilities and expectations.

Bayesian statisticians often point out that their methods "work" and this proves the scientific value of the Bayesian theory. This is a statement about the methods and the choice of prior distributions. It is obvious that if prior distributions had been chosen in a considerably different way, the results would not have been equally impressive. Hence, prior distributions have hardly the status of arbitrary opinions. They are subjective only because much personal effort went into their creation.

One could say that (some) Bayesian statisticians are victims of the frequency statisticians' propaganda. They believe in the criticism of Bayesian statistics that subjectivity and science do not mix. As a reaction to this criticism, they try to justify their usage of subjective priors by invoking de Finetti's philosophical theory. In fact, using subjective priors is just fine because the whole science is subjective in the same sense as subjective priors are. Science is about matching idealized theories with the real world; the match is necessarily imperfect and subjective. Bayesian priors are no more subjective than, for example, assumptions made by physicists explicitly and implicitly in the Big Bang theory. The only thing that matters in all sciences, including statistics, is the quality of predictions.

PART 3
Sociology of Ideologies

Chapter 13

Hypocrisy++

The reader must have realized by now that I consider the subjective philosophy of probability to be one of the greatest intellectual failures of all time. Immediately, a natural question springs to mind—why did a completely irrational philosophical theory acquire a respectable status in Bayesian statistics? I will analyze several general mechanisms that allowed subjectivism to take root in science and popular science.

I have implicitly suggested that irrational theories are acceptable inside the domain of philosophy. I indeed believe that this is true. Respectful analysis of irrational theories does not make philosophers irrational. If philosophers want to be reasonable and rational, they must show respect for and provide a full and honest analysis of totally irrational theories. Doing so allows them to test philosophical arguments and brings to light subtle points.

The titles of the book and this chapter, "Hypocrisy++," refer to a very effective mechanism used by ideologies to hide extremely inconvenient facts. Intriguingly, hypocrisy++ arose in the two leading philosophical theories of probability. This mechanism is not limited to probability; I will show how it was applied in communism and Catholicism. Some of the cases of hypocrisy++ appear within currently popular ideologies. I will not wade into current culture wars to not distract the reader from my main concerns with the philosophy of probability. But I invite the reader to find examples of hypocrisy++ in contemporary ideologies and politics.

The second propaganda mechanism used by some ideologies, including the subjective and frequency philosophies of probability, is example censorship (see Chap. 14). Textbooks and monographs on some philosophical topics fail to discuss natural, obvious and simple examples because these are inconvenient. Many of these examples are capable of destroying a theory. It is amazing how effective the silence is. If you do not talk about a problem, it does not exist. The same technique works very well in mass media, politics, religion, etc.

In Chap. 15, I will discuss Insular Ideological Communities. I will list many mechanisms that allow them to function and harbor even the most bizarre ideas.

13.1 Hypocrisy++

According to the Cambridge Dictionary, hypocrisy is "a situation in which someone pretends to believe something that they do not really believe, or that is the opposite of what they do or say at another time." Normally, hypocrisy is accompanied by a sense of shame. Hypocrites typically try to conceal their hypocritical behavior.

"Hypocrisy++" is a situation when the hypocritical behavior or its consequences are proudly displayed. In a nutshell, some actions or their consequences are so conspicuous that one cannot possibly hide them. So, instead of trying to hide what cannot be hidden, it is best to display it with pride. I will start the chapter with examples illustrating the difference between standard hypocrisy and hypocrisy++.

The term "hypocrisy++" is inspired by the name C++ of a programming language. The meaning of "++" in the programming language is "increment by 1."

13.2 Catholic Church

13.2.1 *Hypocrisy*

The Catholic Church is known for its very strict teaching on sexuality. Only married couples can (and should!) have sex. There are no exceptions to this rule under any circumstances. On top of that, even married couples can use only one contraception method, the so-called "natural family planning" which relies on abstinence during the woman's fertile phase.

In recent decades, many scandals shook the Catholic Church in several countries when the protection of pedophile priests by the church was brought to light. These practices illustrate the standard hypocritical behavior—an attempt to hide actions violating the professed principles.

13.2.2 *Hypocrisy++*

I will discuss two cases of Catholic hypocrisy++. The first is better known, but the second one is where the essence of hypocrisy++ is more apparent.

The conquest of the part of the Americas nowadays referred to as Latin America was unprovoked and brutal. The most advanced American civilizations in existence at that time, the Aztecs and Incas, were much too weak to pose a real threat to Europeans in the foreseeable (at that time) future. Most of the early deaths among Native Americans are attributed to diseases brought by the Europeans to the new continent. However, the subsequent enslavement of the local populations, land grabs, exploitation of natural resources such as gold and silver, etc., constituted one of the most brutal treatments of any population in human history. This was accompanied by the destruction of much of the local cultures and forcible conversion to the Catholic faith.

Blaming the Catholic Church for those historical events would be a great exaggeration. Nevertheless, the Church's role in them is undeniable. At this time, Latin America might be the main bastion of the Catholic Church worldwide because religion lost much of its significance and power in Europe. The Catholic Church displays little shame for how its reach was extended to the American continent.

The second case of Catholic hypocrisy++ occurred in Poland, my native country. In 966, Mieszko I, the ruler of Poland, converted to Christianity. He ordered all Polish people to convert to Christianity as well. According to [71],

> [...] most of the Polish population remained pagan until the pagan reaction during the 1030s.

According to another Wikipedia article, [72],

> The pagan reaction in Poland was a series of events in the Kingdom of Poland in the 1030s that culminated in a popular uprising or rebellion, or possibly a series of these, that destabilized the Kingdom of Poland. [...] Dissatisfaction with the process of Christianization, which had started after the baptism of Poland in 966 was one of the factors that led to the uprising. The Roman Catholic Church in Poland sustained substantial losses, with many churches and monasteries destroyed and priests killed.

The pagan uprising that occurred 60 years, or about two or three generations, after the baptism of Poland proves (if anybody needs proof) that the population was forcibly converted to Christianity.

In 1966, the Polish Catholic Church celebrated the Millennium of the Baptism of Poland with the greatest pomp possible under the communist regime.

The forcible conversion of the Polish population to Christianity can hardly be reconciled with the Biblical principle "thou shalt love thy neighbor as thyself." Yet I doubt that anybody, in Poland or outside, in the twentieth century shed a tear for the ancient pagans forced to abandon their religion. The reason is simple: the victors write history books. At this time, Poland is almost entirely Catholic, and the last pagan must have died more than nine hundred years ago.

I wonder if this case of hypocrisy++ should earn another plus and be known as hypocrisy+++. The celebration of the forcible conversion of the Polish population from one ideology (paganism) to another ideology (Christianity) for political reasons 1,000 years ago was happening at the time when another case of forcible conversion of the Polish population from one ideology (Christianity) to another ideology (communism) for political reasons was going on in parallel. Poland was a communist country from 1945 to 1989. Throughout the twentieth century, communism appeared to be the most progressive political system to many people. It had impressive successes, from political (revolutions in Russia and China), military (victory over Germany), and propaganda (communist movements in Latin America, Asia and Africa) to scientific and engineering (the first satellite and the first man in space). In 1966, Communists tried to align Poland with the new emerging global ideology, just as Mieszko I, the Polish prince, tried to align Poland with the new emerging global ideology in 966. The similarity and irony were lost on the Catholic Church in 1966, protesting the violation of religious freedom.

I add parenthetically that my application of the modern concepts of "Poland" and "Polish nation" to events one thousand years ago is a gross oversimplification. Still, it does not affect my main point about hypocrisy++.

The example of hypocrisy++ concerned with celebrating the Millennium of Baptism of Poland in 1966 is written from the current perspective. According to the currently popular "progressive" ideology, the forcible conversion of the whole population to Christianity was a case of "cultural genocide." However, in 1966, the attitudes towards Indigenous cultures worldwide (for example, in the US, Canada and Australia) were different. At that time, the baptism could be considered a transformation of uncivilized Polish tribes into a civilized nation.

13.3 Communism

13.3.1 *Hypocrisy*

One of the basic tenets of communism is economic equality. This supposedly extends to the people ruling communist countries. From the earliest times of communism in Russia to the current most extreme example of a communist country, North Korea, the leaders enjoyed lives much more luxurious than the rest of the society, even if they did not reach the status of Western billionaires.

The privileges of the top leaders of communist countries and those of communist party apparatchiks were always scrupulously hidden from the public eye. Communists knew they were breaking their principles and behaved like any standard hypocrite—they tried to hide their hypocritical behavior.

13.3.2 *Hypocrisy++*

I grew up in a communist country, so I was indoctrinated in the communist ideology, like all other students. At the college level, we were taught Marxist philosophy, Marxist economics and Marxist political science at a level much more sophisticated than the vulgar newspaper propaganda. Marxism was presented as a scientific theory consisting of objective and unchangeable laws of social evolution. One of these laws said that there was a strict match between the stage of the development of the means of production and the social, political and ideological structure, with the latter elements of the civilization determined by the means of production. Humanity was going through consecutive stages: slavery, feudalism, capitalism, and, eventually, communism. Revolutions were unavoidable because new means of production caused societal tensions that could not be resolved in any way except for a revolution. Early communists believed that the first communist revolution would occur in one of the most developed countries, for example, Germany.

There were only two independent and truly significant

communist revolutions in the world—those in Russia and China. Both countries were in the pre-industrial stage at the time of their revolutions. Communists set out to industrialize both countries and succeeded after some time.

The industrialization of the Soviet Union and China has always been a source of great pride for these countries. It formed a part of the core of the official propaganda, contrasting the prerevolutionary poverty with the postrevolutionary abundance (according to the communists) due to rapid industrialization.

Hypocrisy++ was exemplified by teaching at colleges the old fossilized canons of the Marxist ideology that included the strict match between industrialization and the capitalist political system, and bragging about industrialization implemented by communists. According to Marxism, industrial workers were supposed to bring about the communist revolution. In reality, communist revolutions created a class of industrial workers in the Soviet Union and China. The two revolutions and their aftermath falsified the existence of a supposedly unshakable and scientific link between the stage of development of the means of production and the governing political and ideological system.

Only a few theoreticians were bothered by the discrepancy. Both the communist rulers and the populations enjoyed industrialization. At that time, ecological concerns were much less prominent than nowadays. Rapid industrialization improved the standard of living for the masses and supported the military, which governments needed to maintain power.

13.4 Hypocrisy++ in subjective philosophy of probability

The standard form of hypocrisy in the subjectivist ideology consists of enforcing probability values like the values of any other objective scientific quantity while insisting on probability's totally subjective character. I illustrate this claim with extensive quotes from [7] in Sec. 16.3.2-Sec. 16.3.3.

De Finetti is best known for two contributions to probability. One of them is the subjective philosophical theory of probability. The other one is the theoretical concept of "exchangeability."

A widespread opinion among subjectivists and Bayesian statisticians is that exchangeability is one of the greatest achievements of theoretical probability and statistics and their applications. I agree.

De Finetti's philosophical theory implies that exchangeability is useless because exchangeability is a form of symmetry, and de Finetti dismissed symmetry as a means of finding objective probability values. He did not advance any argument that some consistent probability distributions are more beneficial than others. The essence of subjectivist hypocrisy++ is the effusive admiration of the concept of exchangeability that flagrantly violates the tenets of the subjectivist philosophy.

The basic idea of the subjective philosophy of probability invented by de Finetti is that there is no objective probability. Hence, probability values people use cannot be right or wrong, close to or far from the truth. De Finetti argued that subjective probabilities should satisfy the axioms of a measure, but this is the only rational restriction on their values. De Finetti explicitly argued that symmetry could not be used to find true probability values. According to de Finetti, symmetry is a personal opinion, so it cannot be used to find universally valid probability assignments.

Exchangeability is a form of symmetry. Random variables $X_1, X_2, \ldots, X_n$ are called "exchangeable" if any possible sequence of their values $x_1, x_2, \ldots, x_n$ is as likely as any permutation $x_{\pi(1)}, x_{\pi(2)}, \ldots, x_{\pi(n)}$of these values (here π represents any permutation of $(1, 2, \ldots, n)$; a permutation is a reordering of a sequence). For example, $(3, 5, 7, 2, 8, 1)$ and $(1, 2, 3, 8, 7, 5)$ are equally likely to be observed if the random sequence is exchangeable.

According to de Finetti's philosophy, there are no rational reasons to use exchangeable probabilities in real-life situations. Every consistent probability distribution eliminates all Dutch books.

Let us consider potential secondary reasons for the use of exchangeability. Exchangeable probability distributions are simple but far from the simplest. Suppose you are presented with a deformed coin, and you want to form a probabilistic opinion about the outcomes of the tosses. The standard statistical model for experiments of this type is that the outcomes are exchangeable with the uniform prior distribution. Due to de Finetti's theorem, this may be rephrased as saying that the tosses are i.i.d. (independent with the same probability of heads), with the probability of heads on a single toss being an unknown (random) number, distributed uniformly between 0 and 1. For a simpler model, assume that the tosses are i.i.d. with a known probability of heads, say, 0.3. This model is simpler because the Bayesian updating of the opinion required after making observations of some tosses is more straightforward in the latter case. Both coin-tossing models discussed in this paragraph are classical and simple. Many would argue that the difference between "exchangeable" and "i.i.d." models is negligible regarding the effort needed to apply them in practice. This is true, but my example is meant to be an accessible presentation of a much stronger claim—replacing "exchangeable" models with "i.i.d." models would save millions of dollars. An enormous amount of money is spent on Bayesian statistics for salaries of statisticians and experts using statistics, hardware and software. Almost all of these expenses would be eliminated if, in every case of Bayesian analysis, the "exchangeable" model were replaced with the "i.i.d." model. On the technical side, that would mean applying an atomic prior distribution.

Consider another simple example. Suppose a statistician was asked to analyze two sets of data, one coming from physics experiments on electrons and another from an agricultural station studying a new fertilizer. It may be that the statistician applied an exchangeable model to each data set. Depending on the data structure, it is possible that an exchangeable model can be applied to the combined data from both experiments. This would be the case if all data were "zero or one" type of data, recording whether some event occurred. No scientist or statistician would ever combine the two data sets into one and apply the exchangeable model to them. It is evident to every scientist that some data sets are truly or approximately exchangeable, and some are not.

The final possibility that I will consider is pure "coincidence." It is theoretically possible that all scientists like exchangeability for some "psychological" reasons that have nothing to do with its correspondence to reality. This is explicitly suggested by de Finetti in the following quote (after [9], p. 70):

> Our point of view remains in all cases the same: to show that there are rather profound psychological reasons which make the exact or approximate agreement that is observed between the opinions of different individuals very natural, but there are no reasons, rational, positive, or metaphysical, that can give this fact any meaning beyond that of a simple agreement of subjective opinions.

I do not believe that a watertight logical argument can disprove the claim of "simple agreement of subjective opinions." But of all possible explanations for the universal adoption of exchangeable models, coincidence must be the weakest. One can label every regularity in human activity as a coincidence. Coincidence can explain everything; hence, it explains nothing.

13.5 Hypocrisy++ in frequency philosophy of probability

Whenever we talk about hypocrisy, we implicitly assume that hypocrisy is an act committed or an attitude held by a person or an organization. I am about to attribute hypocrisy to a philosophical theory. This is because, despite the considerable popularity of frequency statistics, it would be hard to find a statistician or a scientist holding views even remotely close to von Mises' radical theory.

The standard form of hypocrisy in the frequency ideology is the practice of using and enforcing probability values of single events that do not fit into any collective (in other words, an i.i.d. or exchangeable sequence), and at the same time insisting that probability can be identified with frequency.

The main philosophical claim of von Mises, mostly forgotten

or ignored, is that a single event does not have a probability, no matter how it relates to other events. Even if experiments or observations are repeated many times, and one can easily determine the long-run frequency of an event, a single instance of the event does not have a probability, according to von Mises.

Hypothesis testing is the flagship of all classical (frequency) statistical methods. I doubt that it was chosen to play this role by frequency statisticians themselves. Paradoxically, Bayesian statisticians have elevated hypothesis testing to this role. It is possible that Bayesian statisticians do not like any frequency statistical methods but reserve their extreme wrath for hypothesis testing. I will discuss a case of this controversy illustrating the prominent position of hypothesis testing, deserved or not, within the frequency ideology in Sec. 16.2.10.

Hypothesis testing can be applied in many practical contexts, such as medicine or quality control in manufacturing. I will focus on hypothesis testing in science. Much of scientific research fits the following scheme. A scientist chooses a project and collects some data. Then, she analyzes the data. This step may involve testing a hypothesis or a few hypotheses. The results are published, and the scientist moves on to the next project. The new project is often focused on a different natural phenomenon, the data have a different structure than in the previous project, the amount of data is different, and the test statistic is different. The two statistical analysis cases only have a significance level in common. In other words, the probability of a false positive is set at the same level. This does not have to be the same in all instances, but it is popular to set the probability of a false positive at 5% or 1%.

An implicit (and only, as far as I can tell) justification for hypothesis testing in the context of the frequency philosophy is that, although there is no guarantee that any case of hypothesis testing will result in the correct decision to reject or accept the null hypothesis, in the long run, the percentage of false positives will be small (5% or 1%, depending on the significance level) among those cases when the null hypothesis is true. This interpretation of hypothesis testing is a clear case of hypocrisy++. According to von Mises, probability can be applied only to collectives, that is long sequences of identical experiments or observations. All relevant characteristics of these experiments or observations have to be identical—except probability, because, according to von Mises, individual elements of the sequence are not associated with any probability. Individual tests in a sequence of scientific hypotheses tests have nothing in common, so they do not form a collective. The only thing they may have in common is the significance level, but, according to von Mises, this is precisely what they do not have—the concept of probability does not apply to a single instance of false positive. Hypothesis testing is the photo-negative of von Mises' philosophical theory.

13.6 Art: Non-hypocrisy++

The article [73] reports that "In December 2004, Duchamp's *Fountain* was voted the most influential artwork of the 20th century by 500 selected British art world professionals." The ultimate piece of modern art was a ready-made urinal displayed at an exhibition in 1917, lacking any connection with the traditional idea of art. *Fountain* was not made by the artist, had no particular aesthetic appeal, and had no meaning except for the obvious (scatological, offensive). Its main and only function as a piece of art was to offend the viewers and make them think about the essence and meaning of art.

Hypocrisy++ can occur only within a rigid ideology. Art does not have such an ideology. Every new trend in modern art was accompanied by a manifesto supplied by an artist or a group of artists or was supported by art critics doing their penitence after failing to recognize the greatness of van Gogh within his lifetime. So, no, declaring Duchamp's *Fountain* the greatest piece of art of the 20th century was not a case of hypocrisy++.

But the declaration of the British critics had some attributes of hypocrisy++. Over the 20th century, modern art pushed the limits, and critics tried to stay abreast of the changing trends and sensitivities, effectively abandoning all quality criteria. They painted themselves into the corner. They either could declare Duchamp's *Fountain* a hoax and admit that they had been fooled or go to the other extreme and put the urinal on the highest pedestal. The first option was unattractive because it would have opened the floodgates—much of the modern art would have to be reevaluated and demoted. Museums would look silly in retrospect. Art criticism was not blemished by hypocrisy++ but could not escape the workings of its inner mechanism.

13.7 The audacity of hypocrisy++

Hypocrisy++ works. Very few people remember, realize or care that Poland is Christian as a result of highly unchristian behavior, that the industrialization of the Soviet Union and China contradicted Marxist prophecies, that applications of exchangeability contradict the fundamental tenets of the subjective philosophy of probability, or that hypothesis testing is the photonegative of the frequency philosophy of probability.

Hypocrisy++ occurs for several reasons. The first, illustrated by the examples of the conquest of Latin America and the industrialization of the Soviet Union and China, is that some actions or consequences are so conspicuous that one cannot hide them. So, instead of trying to hide what cannot be hidden, it is best to display it with pride.

Second, some hypocritical actions appear to have positive consequences from the current perspective. One can only imagine what pagan Poles thought about the forcible Christianization of Poland. However, the current Polish population is primarily Christian, so their perspective is entirely different—to them, Christianization was a positive historical development. Similarly, few ordinary people in the Soviet Union and China would be against industrialization, whether it agreed with the communist tenets or not.

Next, audacious hypocritical behavior is unlikely to be recognized as such. People seem to hold a natural suspicion of all things deliberately hidden by others. Hiding an object or an action signals a possible violation of some rules. False logic implies that items or actions that are conspicuously paraded agree with the rules. I do not know whether these human abilities and attitudes are due to "nature" or "nurture"; I would guess both.

The herd instinct helps hypocrisy++. Almost all Bayesian statisticians believe that exchangeability agrees perfectly with subjectivism not because they studied the philosophy of probability but because "everybody knows" that this is the case. The same applies to hypothesis testing. I doubt that any of the two groups of people, those who support hypothesis testing and those who criticize it, spent much time studying von Mises' philosophy. Hypothesis testing is widely considered part of the "frequency" ideology because this is the accepted wisdom.

Chapter 14

Missing Schrödinger's Cat

Einstein discovered the law $E = mc^2$. If you run out of gas but have a cheese sandwich, you can turn the sandwich's mass into energy that will power your car until you reach the nearest gas station. Do you believe that my example is a correct and realistic implementation of Einstein's theory?

Examples are like air. They are obviously there, and we cannot live without them. We take them for granted. Most people do not spend any time on their theoretical analysis.

We learn by observing many examples or doing many experiments. Our understanding of the theory uniting our experiences comes later. Both Kuhn in [56, pp. 187–191] and Polanyi in [55] point out that the learning process is based at least as much on reviewing many examples or practicing a skill, as it is on studying the theoretical component of a field of knowledge.

According to [74], "In quantum mechanics, Schrödinger's cat is a thought experiment that illustrates a paradox of quantum superposition. In the thought experiment, a hypothetical cat may be considered simultaneously both alive and dead [...]." This thought experiment exposes a deeply counterintuitive consequence of quantum mechanics. The experiment seems to contradict a seemingly obvious observation that a cat cannot be dead and alive simultaneously. This attack on quantum mechanics stimulated a lot of literature and helped clarify some aspects of theoretical quantum mechanics and its applications.

Science does not fear thought experiments that challenge the ruling theories. The analysis of these examples is needed; one could even say that it is necessary. The only result of this analysis may be an improvement of scientific theories.

I recall an interview with Noam Chomsky in which he said his TV interviews were unsuccessful because his views were nonstandard. A short interview is sufficient if the interviewee presents only popular beliefs. Typical TV interviews were not sufficiently long enough for Chomsky to explain his nonstandard opinions. For similar reasons, radical philosophical ideas require diverse examples to become comprehensible.

I will argue that some ideologies avoid discussing obvious examples or thought experiments because they could lead to embarrassing conclusions and undermine the theories.

This chapter contains my own examples, my own analysis of these examples, and some general ideas. It is important to remember my main message—these are examples that are missing from the literature. I cannot expect philosophers to have stated and discussed examples identical to mine. I am missing a clear understanding of what subjectivists and frequentists would have answered to challenges presented by my examples. I might have agreed or disagreed with their analysis. At this time, I cannot do either because I cannot find a clear analysis of examples similar to mine in the existing literature.

However, I will start on a positive note.

14.1 Education

The significance of a concept often derives from its ubiquity. When I teach calculus for first-year students, the concept of integration is illustrated by its applications to area, volume, work, and center of mass. This is just the tip of the iceberg. Undergraduate students who major in STEM (science, technology, engineering and mathematics) will soon learn many more applications of integrals. Students who go on to graduate school will discover that integrals are indispensable in fields as diverse as biology, medicine, economics, sociology and psychology (or at least in some subareas of these fields).

14.2 Popular science

Examples play a major role in communication between scientists and the general population. Modern physics is almost incomprehensible to most people. Only a tiny fraction of humanity can reasonably understand Schrödinger's equation (in the foundations of quantum mechanics) or Einstein's equations for general relativity. I am not among this tiny fraction. The only way a physicist can explain the fundamental ideas of quantum and relativity theories to a wide audience is through examples. These include the double-slit experiment (see [64]), Schrödinger's cat (see [65]) and the twin paradox (see [75]). Some more specialized examples and ideas include the EPR paradox (see [76]) and Bell's inequalities (see [76]). Black holes and Big Bang can also be included on this list.

14.3 Art

Art is a strange illustration of the significance of examples. On the one hand, museums and galleries display numerous art pieces accessible to all who have enough money and time to reach the museums (many museums in the developed world are free). On the other hand, many modern art pieces have no features distinguishing them from wallpaper or random objects.

I once visited a museum with someone without prior exposure to museums or fine art. I was surprised that he could not see considerable differences between the styles of the two paintings hung side by side by Zurbarán and El Greco. The latter painter is one of my favorites. I learned to recognize his paintings by viewing many examples.

I love modern art. Some of my favorite artists are Picasso, Miró, Giacometti and Botero. I can easily recognize their art. On the other hand, I believe museums harbor many pieces considered art only because museums say so. Those pieces have no distinguishing characteristics from ordinary objects and could be easily imitated (and forged if anyone wanted to forge them).

I challenge the reader in Fig. 14.1. Can you tell which of the four pictures is high art? I am sure that this quiz is very easy for art connoisseurs. At the same time, I think that the quiz is practically impossible to solve for a large segment of the general public.

Fig. 14.1 All four images were digitally generated. Some images are (scaled) copies of paintings considered high art, that is, items displayed in museums or sold by high-end galleries. Can you guess which of these images represent fine art? See the end of Sec. 15.1.4 for the answer.

What art is missing are clear examples and explanations of why some pieces of art similar to those in Fig. 14.1 are considered to be high art, why some of these paintings are very expensive, and why some of them are displayed in museums, while paintings made by ordinary people in the same style would not enjoy even one percent of that recognition.

The reader must have noticed that I have chosen very simple (minimalist) pictures for my test in Fig. 14.1. Most of the modern art is more complex than that.

Art provides an excellent illustration of the significance of examples. On the one hand, one can learn to recognize and appreciate different artists' styles by viewing many examples. On the other hand, widespread suspicion of the value of modern art can be attributed to the lack of examples illustrating the selection criteria for the inclusion of radically minimalist art pieces in museum collections.

14.4 Philosophy

One reason philosophy has a bad reputation among some scientists and laypeople is that some of the best-known and radical philosophical claims come with clearly tendentious examples or no examples.

14.4.1 *Motion does not exist*

According to Zeno, motion does not exist. He based his claim on the analysis of some examples, including Achilles chasing a tortoise. The choice of tortoise among all possible animals that Achilles could have chased was always a mystery to me, but there might have been good reasons for this choice in ancient Greek culture.

The striking feature of this ancient philosophical claim is that it does not discuss its practical consequences illustrated by examples from ordinary life. Horses were known to the ancient Greeks. Normally, one would presume that a motionless horse was dead and a moving horse was alive. This entailed predictions that the dead horse would never move (by itself) and the alive horse would stay alive and do some work, at least for some time. I am unaware of any discussion by Zeno of examples of this type—what should one think about motionless and moving horses? Since motion does not exist, should we abandon the illusion that one of the horses is moving? Should we stop using horses for travel and transportation? Or should we beat a dead horse until it moves?

14.4.2 *Theology*

I grew up as a Catholic in a mostly Catholic country, so I will limit my discussion of religion to Catholicism. Other religious philosophies may suffer from problems similar to those presented in this section, but I do not want to venture into unknown waters.

All events that occur in our universe support the infinite benevolence of God, according to Catholic theologians. Catholic or not, all people can easily understand why every happy ending could be evidence of God's infinite love. However, Catholics do not stop here. The most extreme misfortunes (short of death) that the Old Testament writers could think of were piled on Job (see [77]). The Holocaust seems to be the most extreme recent example of human misery. According to Catholic theologians, neither the story of Job nor the Holocaust makes a dent in the theory of God's infinite benevolence.

In theology, the problem is not that examples are missing. The problem is that they are not helpful to non-believers. Examples are supposed to illustrate a claim by showing when the claim holds and when it does not. Catholic theologians present God's omnibenevolence as an unshakable fact. Examples with happy endings are misleading because they implicitly suggest that some examples may lead to the opposite conclusion.

It is possible that ordinary Catholics adopted the philosophical stance of Catholic theologians. It is more likely that Catholic theologians formalized the beliefs of ordinary Catholics who want to believe in the benevolent God no matter what the empirical world may be.

I like medieval churches. Their frescoes show hell and the devils or the Beast with great emotion and power. This religious and artistic trend stopped in the Renaissance and signified a profound change in the Catholic religion. I always found the depictions of heaven lacking. Medieval paintings often show static worshipers of God. Saints sitting on clouds and playing harps are the modern cliche. The vivid depiction of hell and the lack of explicit examples of heavenly activities represent our culture. We know various ways people can be made to suffer, which were extensively practiced throughout history. Many activities people consider pleasurable or entertaining would not be appropriate in the divine presence. An implicit message is that the faithful should trust God (or the Catholic Church)—their righteous lives will be rewarded appropriately.

14.4.3 *Marxism*

Marxism, when it comes to examples, is just like the Catholic religion. There is a striking imbalance between negative and positive examples but it takes a different form. Negative examples, primarily accounts of actual events and situations, are numerous and deeply resonating. Various writers, not necessarily Marxists, documented the suffering of the working class.

Marx painted a vision of communism based on the principle "From each according to his ability, to each according to his needs." I grew up in a communist country, and I was indoctrinated in Marxism at college. I do not recall even a single example illustrating Marx's idea. If five million Polish people need helicopters, would each get a private helicopter when the country becomes fully communist?

14.4.4 *Relativism*

According to [30],

> Relativism, roughly put, is the view that truth and falsity, right and wrong, standards of reasoning, and procedures of justification are products of differing conventions and frameworks of assessment and that their authority is confined to the context giving rise to them. [...] Relativists characteristically insist, furthermore, that if something is only relatively so, then there can be no framework-independent vantage point from which the matter of whether the thing in question is so can be established.

The cited article contains many examples. Since I am complaining about the inadequate supply of examples in this chapter, I will discuss how this problem affects two specific publications.

Margolis makes it clear in [78] that he interprets relativism as a rejection of *tertium non datur* (the law of excluded middle). As far as I can tell, the book contains only two half-hearted examples. Without examples, his "baroque analytical philosophy" (Margolis says that someone characterized his philosophy this way) is no more than an intricate wordplay. I have no clue how to apply his theory in simple or complex situations; therefore, I have no clue whether I agree or disagree with his philosophical theory.

Examples given by Feyerabend in [31] seem to be designed only to provoke and annoy, for example, the author's suggestions that it is safe to jump from a skyscraper and that superstition should be taught at school in parallel to science. An extensive portfolio of similar examples illustrating everyday life, business, government and military operations would make relativism a laughing stock.

14.5 Subjective philosophy of probability

A striking feature of the subjective philosophy of probability is the lack of examples illustrating several key concepts and obvious questions. There are almost no (good) examples illustrating the Dutch book argument, the role of the prior distribution, unusual consistent probability distributions, and unusual applications of exchangeability.

14.5.1 *Dutch book*

Recall that a "Dutch book" is formed against a person if she makes some decisions in such a way that no matter which events occur and which do not occur, she will lose some money. A Dutch book can be formed against a physical person but also against a family, army, company, university, country, etc. In general, a Dutch book can be formed against an agent or entity who, as a whole, can gain or lose money or, in general, utility (for example, friendship or respect). One should note that the decision-maker need not be the same as the agent who found themselves in the Dutch book situation. For example, an employee of a company could make decisions that would put the company in a Dutch book. A doctor could make decisions that could put a patient in a Dutch book.

All published examples of a Dutch book fall into three categories.

1. Artificial situations resembling psychological experiments.
2. Betting or gambling.
3. Arbitrage pricing theory.

Artificial situations are just that—artificial. One can imagine and write about such situations; in principle, one can run psychological tests of this type. The main problem with this application of the concept of a Dutch book is that, by the definition of "artificial," there is no good reason why any of these situations should occur. If anyone finds themselves in a Dutch book in an artificially designed environment, it is only because that was the experimenter's intention (with, presumably, the subject's consent).

Before discussing gambling situations, the primary examples in the literature, let us quickly look at arbitrage pricing (see [79]). This theory of pricing "contingent claims," of which the "European call" option is best known, is a highly sophisticated mathematical theory based on several assumptions and applicable to a narrow sliver of investing activity. The central concept of this theory, "arbitrage," is a situation in which an investor can make money without risk. Hence, it is the photo-negative of the concept of a Dutch book. There are significant differences between arbitrage pricing theory and de Finetti's philosophical theory, but I will limit myself to pointing out just one of them. The most profound difference is that, under some assumptions, the arbitrage pricing theory says that a rational investor must use a uniquely identified consistent probability distribution, known as the "equivalent martingale measure" in this context. This is exactly the opposite of what de Finetti claimed—that there is no unique rational, consistent probability.

There is no question that people placing sports bets can find themselves in a Dutch book situation. A typical bookmaker offers bets on different outcomes of a sports event. The odds on different outcomes are slightly "unfair" to the player—this is justified by the fact that "bookmakers" are companies that must generate profits. If a player places several bets at once, it is easy to see that some combinations and amounts of bets will generate a Dutch book. Statistical analysis of Dutch books related to gambling is conspicuously missing from the literature on the foundations of probability. Since Dutch books can be created, how prevalent are they? How much money is lost due to Dutch books? What percentage of bookmakers' profits can be attributed to Dutch books?

Probability is used in various components of the social structure. Let us set aside the mathematical theory of probability and statistics, both theoretical and applied. Science (physics, biology, economics, etc.), medicine, manufacturing, trade, investments, financial markets, insurance, military, and government all involve risks, hence, probability. The striking feature of the philosophical literature on probability is that it lacks any examples illustrating the Dutch book idea in any of these contexts. This void is not because philosophers are not smart. The reason is much more fundamental: the Dutch book is a red herring. Philosophers did not present realistic examples of Dutch books because there are none.

14.5.1.1 *Medicine*

Let us try to find a Dutch book in some areas of practical life. First, let us have a look at medicine. Suppose a doctor and a patient have to decide which treatment is best for the patient: chemotherapy, radiotherapy or surgery. All treatments typically carry some benefits and risks, including side effects. None of these can be predicted with certainty, so any decision involves probability. How the patient or the doctor (or a hospital or an insurance company, etc.) can find themselves in a Dutch book situation? Which combination of the three available treatments will deteriorate the patient's condition with absolute certainty? Which combination of the three available treatments will result in a loss of utility for one of the agents (patient, doctor, hospital, etc.) with certainty? None, obviously.

14.5.1.2 *Investments*

Imagine a US car company considering building a new car assembly plant in Mexico. The investment may be profitable or money-losing. None of the two possible outcomes can be predicted with certainty, so the investment decision must be based on probability. Which of the two decisions (build the plant or not) could put the company in a Dutch book situation?

14.5.1.3 *Military operations*

In the military context, one can easily spot an opportunity for a Dutch book. The following strategy would result in a sure loss of utility—gather all of your troops in one place and explode a bomb in their midst so that they are all killed. Is this strategy what the military is trying to prevent by using probability?

In a more realistic scenario, a commander has to choose the time, place and manner of attack on the enemy. The outcome cannot be predicted with certainty, so probability is a part of the decision problem. I do not see how a suboptimal decision can create a Dutch book.

14.5.1.4 *Useless information*

I am not aware of any discussion of the following very common scenario. Suppose that a decision-maker is preparing to make some decisions. Before she finishes her analysis, some new information becomes available. She has not made any decisions yet, so a Dutch book cannot arise from the "past" decisions. It follows that she can completely ignore the latest piece of information. She can eliminate Dutch books whether she includes the new information in her calculations or not. My abstract example represents actual Bayesian statistics. Users of Bayesian statistics make no decisions before the data analysis is complete.

It is remarkable that, to this day, the Dutch book argument is sympathetically presented in some books without discussing real-life examples, and some readers seem to accept a theory disconnected from reality.

14.5.2 *Prior distribution*

The concept of prior distribution illustrates a problem that appeared in philosophy long ago. A highly abstract argument is sometimes accepted as proof of a common belief despite the lack of a discernible relationship between the two.

To illustrate my claim, let us recall one of the proofs of God's existence by St. Thomas Aquinas (see [45]):

> In the world, we can see that things are caused. But it is not possible for something to be the cause of itself because this would entail that it exists prior to itself, which is a contradiction. If that by which it is caused is itself caused, then it too must have a cause. But this cannot be an infinitely long chain, so, there must be a cause which is not itself caused by anything further. This everyone understands to be God.

It is a great mystery how this abstract proof can be linked to God, who created Adam and Eve.

Going back to probability, consider prior distributions in statistics. Let us start with the role of the prior distribution in de Finetti's theory. According to de Finetti, coordinating decisions is the only practical application of probability. A decision-maker uses a consistent family of probabilistic beliefs to make decisions until new information is available. Then, the decision-maker should apply the Bayes theorem to arrive at a new consistent family of probabilistic beliefs, once again, to make rational decisions. It follows that the prior distribution is useless and can be safely abandoned if no decisions are made before the information arrives.

The presumed link between de Finetti's theory and prior distributions in statistics is not illustrated by any examples in the literature. It is common to choose a prior distribution after collecting the data. At first sight, this makes no sense because one cannot make a prior decision after collecting the data.

One could defend a delayed choice of the prior by saying that the prior distribution summarizes the guiding principles used in decision making before collecting the data. If this is the case, I am unaware of any examples of how this is implemented in practice. How large is the family of consistent prior distributions compatible with the prior decisions? What criteria should one use to choose a distribution in this family?

Consider the following highly simplified but not unrealistic decision problems linked to a statistical problem. Suppose a pharmaceutical company has patents on Old Drug and New Drug. A new set of data has been collected and analyzed. Before the trials, the company had the following four choices: (i) manufacture none of the drugs, (ii) manufacture the Old Drug, (iii) manufacture the New Drug, and (iv) manufacture both drugs. After the statistical analysis, the company has the same four choices. Which combination of the prior and posterior choices could put the company in a Dutch book situation? Furthermore, how is this related to the prior distribution? How does the prior distribution reflect the prior decision strategy?

Another mystery never explained via examples is the relationship between priors and posteriors. The two distributions represent the decision-maker's beliefs prior to and posterior to the collection of data. The literature strongly suggests that priors are subjective, or at least, they can be subjective and still rational. I do not recall anyone suggesting that statistical posteriors are as subjective as statistical priors. In the writings of subjectivists, Bayesian statistics appears to be a miraculous transmutation of subjective priors into objective posteriors.

14.5.3 *Consistent probability distributions*

Humanity is divided into two parts. Some people do not realize how large the family of consistent probability distributions is because they are not experts in the field. Others do not realize how large the family of consistent probability distributions is because they *are* experts in the field.

Consider the following simple example. A deformed coin will be tossed ten times. The outcomes are unpredictable, so one could adopt the following consistent probabilistic opinion: the results are independent, and the probabilities of heads on the consecutive tosses are

$$\frac{1}{2},\ \frac{1}{3},\ \frac{1}{4},\ 0.001,\ \frac{\sqrt{\pi}}{8},\ \sqrt{0.12345567},\ 0.9,\ 0.99,\ 0.999,\ 10^{-10}.$$

People who do not know any mathematical probability theory cannot tell whether this probability distribution is consistent. People who took an introductory class on probability in high school or college might not be confident enough to judge whether this distribution is consistent. Experts with a Ph.D. must have seen thousands of (consistent) probability distributions in their research, and none of them looked even remotely like our example. Experts would have no problem recognizing that the above distribution is consistent but never consider such distributions. The vast majority of consistent distributions are outside of their radar. The result is that "generic" probability distributions (i.e., distributions chosen "uniformly" from the family of all consistent distributions) are not discussed in the context of the subjective philosophy, although they should be.

Anybody with a solid background in mathematical probability theory can generate an arbitrarily large number of crazy distributions similar to that given above once she absorbs my example. None of the crazy examples is discussed by philosophers sympathetic to the subjective philosophy of probability. Even examples given by the critics of the subjective theory are much more "rational" than the one stated above. The reason for this void in the subjectivist literature is quite obvious. Crazy consistent examples could be classified as perfectly rational. Then the subjective philosophy of probability would be classified as one of the "crazy" directions of philosophy, perfectly fitting among those claiming that there is no motion, no physical world, or no objective truth. While this might not deter philosophers from the subjective theory, its popularity outside philosophy would plummet.

The other option is to explain why "crazy" consistent probability distributions should not be used in practice. However, doing that would destroy de Finetti's philosophical theory. His main philosophical claim is that no objective criterion could discriminate between consistent probabilities. According to de Finetti, none of the consistent distributions is any better than any other.

The spectacular failure of de Finetti's philosophy is papered over simply and effectively—inconvenient examples of crazy consistent distributions are never discussed. This is not limited to thought experiments appropriate for philosophical articles and books. Real-life experiments with "crazy" distributions can be easily performed. For example, one could program a computer to make decisions using an unusual, consistent distribution. The results could be recorded and published. I am unaware of real-life experiments testing decision making using arbitrarily constructed consistent probability distributions.

14.5.3.1 *Conspiracy theories*

My example of a "crazy" consistent distribution given in the previous section is artificial. However, numerous real-life consistent probability distributions are controversial and unusual because a small fraction of society harbors them, and they are often firmly rejected by the majority. Religion, traditional prejudices (racism, antisemitism, etc.), and conspiracy theories are good, although considerably different, examples. We usually do not consider these ideologies probabilistic opinions, but we can express them this way. Perhaps a better way to think about it is to suppose that someone harboring one of these views expresses it in probabilistic terms and demands full respect, citing de Finetti's theory.

Religion is often identified with "faith." This word evokes probability. Many religious people do not claim that their religion's dogmas are empirical or logical truths, but their acceptance requires faith. This faith is rarely blind—most of the time, it is based on holy books, miracles, etc. I will not try to express religious beliefs in a formal, probabilistic way. I think that this would be an easy, although tedious, task.

Racism and antisemitism can be expressed as probabilistic opinions because someone could say that a Black or Jewish person has negative traits with high probability. A similar remark applies to conspiracy theories. For example, one could say that given the available information, it is implausible that man landed on the Moon.

None of these examples is discussed in subjectivist literature. One could argue that none of the religion, prejudices and conspiracy theories are consistent. However, before accepting this argument, I want to see a list of explicit probabilistic inconsistencies. I will consider these ideologies consistent until proven guilty.

Assuming that religion, prejudices and conspiracy theories are consistent, is there a rational reason to reject any of them? No, according to de Finetti's theory. This subjectivist claim is swept under the rug—the probabilistic interpretation of ideologies is never discussed.

Let me point out parallels with relativism. According to this philosophical theory, there is no objective truth. Racism and antisemitism are inconvenient examples for left-leaning relativists.

14.5.3.2 *Money-saving consistent distributions*

Two families of simple, consistent distributions could save much money if one's only goal was to eliminate Dutch books. They are

(i) Distributions concentrated at a single atom.

(ii) Uniform distributions.

A "probability distribution concentrated at a single atom" is a technical (mathematical) way of saying that one is sure of a random experiment or observation's outcome (let us call it O). A great advantage of an atomic distribution over spread-out distributions is that the subjectivist decision strategy is straightforward to implement in this case—all one has to do is to check the value (or utility) of the payoff of a bet in case O occurs. If the payoff is strictly positive, then the bet should be made. Otherwise, it should not. We do not have to invoke de Finetti's theorem to see that a Dutch book cannot be made against the decision-maker who adopted this effectively deterministic worldview. If O happens, then the total payoff from all placed bets will be strictly positive, so there will be no Dutch book.

Uniform distributions often lead to very simple calculations. If you have a deformed coin and plan to toss it n times, you could assume that every outcome, for example, $HTTHTHHTHTH$, has the same probability. The Bayes formula reduces to a trivial calculation in this case. Given any initial sequence of results, the probability of heads on the next toss is exactly 1/2. This example is well-known in the literature on the logical philosophy of probability. In this part of the literature, the uniform probability model is rejected because the model eliminates learning.

I have not seen examples of atomic and uniform distributions discussed in the subjectivist philosophical literature. All philosophical theories of probability, except for subjectivism, reject the two families of distributions. In the subjectivist world, the two families of distributions are in limbo—they are neither explicitly accepted nor explicitly rejected. In this way, subjectivists can avoid answering an inconvenient question.

14.5.3.3 *Probabilistic Dysfunctionality Syndrome*

Dyslexia is a mild disability that makes people misinterpret written words, for example, by rearranging their letters, as in "tow" and "two." Let us consider the case of Mr. P. Di Es, an individual suffering from a probabilistic counterpart of dyslexia, a "Probabilistic Dysfunctionality Syndrome." Mr. P. Di Es cannot recognize events which are physically independent or invariant under symmetries. Hence, Mr. P. Di Es cannot apply (L1)-(L6) to make decisions. Here are some examples of Mr. P. Di Es' perceptions. He thinks that the event that a bird will come to the bird feeder in his yard tomorrow is not physically independent from the event that a new war will break out in Africa next year. At the same time, Mr. P. Di Es does not see any relationship between cloudy skies in the morning and rain in the afternoon. Similarly, Mr. P. Di Es has problems sorting out which sequences are exchangeable. When he reads a newspaper, he thinks that all digits printed in a given issue form an exchangeable sequence, including those in the weather forecast and stock market analysis. Mr. P. Di Es buys bread at a local bakery and is shortchanged by a dishonest baker about 50% of the time. He is unhappy every time he discovers that he was cheated. Still, he does not realize that the sequence of bread purchases in the same bakery can be considered exchangeable so he goes to the bakery with the same trusting attitude every day.

Some mental disabilities are almost miraculously compensated in some other extraordinary ways; for example, some autistic children have exceptional artistic talents. Mr. P. Di Es is similarly talented in an extraordinary way and is absolutely consistent in his opinions, in the sense of de Finetti. A person impaired as severely as Mr. P. Di Es would be as vulnerable as a baby. The ability of Mr. P. Di Es to assign probabilities to events in a consistent way would have no discernible positive effect on his life.

The example is clearly artificial—there are very few, if any, people with this particular combination of disabilities and abilities. This is probably the reason why so many people do not realize that consistency alone is totally useless. Consistency is never applied without (L1)-(L6) in real life. It is amazing that the subjective philosophy, and implicitly the consistency idea, claim all the credit for the unquestionable achievements of the Bayesian statistics.

An example illustrating well the attitude towards consistency in the current science and technology is that of a robot with (some) artificial intelligence (AI) capabilities. It would be easy to program a computer to consistently assign probability values to all events without any reference to real-world symmetries and physical independence. From the current AI perspective, such a program would be laughable. The essence of modern AI is to develop programs that mimic human perceptions and information processing, including recognition of symmetries and independence.

14.5.4 *Exchangeability*

Readers familiar with Bayesian statistics will likely be surprised by my claim that the literature does not contain examples of exchangeable random variables. The concept of exchangeability is fundamental to Bayesian statistics, so it appears in every Bayesian textbook and many real data models.

What I am missing is the philosophical discussion of unusual uses of exchangeability. Suppose a statistician plans to work on two data sets with the same mathematical structure. The first data set represents the results of clinical trials of a new antibiotic tested on patients with pneumonia. The second data set concerns a new fertilizer for tulips. Let $X_1, X_2, \dots, X_{100}$ and $Y_1, Y_2, \dots, Y_{100}$ stand for the (still unobserved) two data sets. Typically, the first data set would be considered exchangeable, and the same would apply to the second data set. But it never happens that the concatenated sequence

$$X_1, X_2, \dots, X_{100}, Y_1, Y_2, \dots, Y_{100}$$

is considered exchangeable. In practice, "exchangeable" means "identical." While different patients may be considered "identical" in an appropriate sense (for example, we cannot tell the difference between them if all we know is their Social Security numbers), no one would consider antibiotic and fertilizer experiments "identical" in any practical sense.

Nevertheless, making the concatenated sequence exchangeable is consistent. If exchangeability is rejected as a scientific model for the concatenated sequence, then there must be a reason other than inconsistency. Subjectivists never discuss examples of unusual applications of exchangeability because they would demolish their claim that there is no objective probability.

14.5.4.1 *Volkswagen emissions scandal*

My example of bringing antibiotic and fertilizer experiments together is artificial, of course. The reader may doubt whether it is worth anybody's time (even philosophers' time) considering such examples. I will now discuss two real-life examples where unusual exchangeability claims were made or implied.

According to [80],

> The Volkswagen emissions scandal [...] began in September 2015, when the United States Environmental Protection Agency (EPA) issued a notice of violation of the Clean Air Act to German automaker Volkswagen Group. The agency had found that Volkswagen had intentionally programmed turbocharged direct injection diesel engines to activate their emissions controls only during laboratory emissions testing which caused the vehicles' NO_x output to meet US standards during regulatory testing, but emit up to 40 times more NO_x in real-world driving.

The data relating to this scandal must have been complex, but the following simplified presentation will likely capture the essence of the incident. Let $X_1, X_2, \ldots, X_{100}$ represent the measurements collected by EPA in the "test mode," and let $Y_1, Y_2, \ldots, Y_{1,000,000}$ denote the hypothetical (but never collected) measurements for the vehicles driven by consumers. Most people would agree that each of the two sequences taken separately was exchangeable. The essence of Volkswagen's cheating was an implicit claim that the concatenated sequence,

$$X_1, X_2, \ldots, X_{100}, Y_1, Y_2, \ldots, Y_{1,000,000}$$

was exchangeable. It is consistent to consider the combined sequence exchangeable. So why does everybody (including the Volkswagen leadership) say the company cheated? Obviously, nobody believes that the combined sequence was exchangeable.

14.5.4.2 *Tobacco Institute*

Wikipedia has this to say about the Tobacco Institute (see [81]):

> The Tobacco Institute was founded in 1958 as a trade association by cigarette manufacturers, who funded it proportionally to each company's sales. It was initially to supplement the work of the Tobacco Industry Research Committee (TIRC), which later became the Council for Tobacco Research. The TIRC work had been limited to attacking scientific studies that put tobacco in a bad light, and the Tobacco Institute had a broader mission to put out good news about tobacco, especially economic news. It also attacked scientific studies, although more by casting doubt on them rather than by rebutting them directly.

This example is a mirror image of the previous one. While in the previous case, the cheating consisted of pretending that a non-exchangeable sequence was exchangeable, in this case, the implicit claim was that some exchangeable sequences were not exchangeable.

The data relating to cigarettes and cancer is much richer and more complex than that related to Volkswagen cars, so there is no way I could present its subtle features (especially since I am not a working statistician). I will propose a toy model that contains the gist of the statistical analysis of the relationship between smoking and cancer.

Let $X_1, X_2, \ldots, X_n$ represent all people living in the past, the present and the future who did not, do not, and will not smoke. Likewise, let $Y_1, Y_2, \ldots, Y_m$ represent all smokers who ever lived and who will live in the future. Let the data be limited to a simple "yes" or "no"—either the person died (will die) of lung cancer or not. According to the simplest statistical model, each sequence is exchangeable. Standard and simple mathematical theorems imply that the frequency of cancer observed in the past will also be observed in the future with a very high probability. Since cancer incidence was higher among smokers, it will also be higher in the future. It follows that decreasing smoking popularity will decrease lung cancer incidence.

The simple statistical analysis presented above is based on the assumption that both sequences are exchangeable. The Tobacco Institute was right that the real data were far from the above clean model. For example, smoking may be related to class status, which may be related to the quality of health care available to smokers. Controlled experiments on humans cannot be conducted because it would be unethical to ask people, even volunteers, to smoke cigarettes for an extended period.

As far as I know, the Tobacco Institute, its critics, and all subjective philosophers of probability missed the crucial point—the two sequences can be declared non-exchangeable. Alternatively, their elements can be declared to be independent. These two probability distributions and numerous other distributions are consistent and quash the presumed correlation between smoking and cancer. Hence, according to de Finetti's theory, the easiest way to proceed for the Tobacco Institute would be to choose a consistent model for the impact of cigarette smoking on health that would suit the tobacco companies. Needless to say, nobody ever believed that one could arbitrarily declare a sequence to be exchangeable or not.

Noam Chomsky is a famous linguist and political activist. He was raised Jewish and, according to [82], "Chomsky had long publicly criticized Nazism, and totalitarianism more generally, but his commitment to freedom of speech led him to defend the right of French historian Robert Faurisson to advocate a position widely characterized as Holocaust denial." Where are principled subjectivists ready to defend Volkswagen and the Tobacco Institute against their personal beliefs but in defense of the general principle?

14.5.5 *Individual freedoms*

Society enforces various rules for various reasons. Here are some of the reasons.

(i) There is a widespread belief in objective reality and objective laws of nature, including science and mathematics. Hence, certain actions' results seem totally or highly predictable. Society enforces rules that benefit the whole society or its subcommunities or individuals. For example, the usual arithmetic is enforced on tax forms. Manufacturers of canned foods cannot add poisonous ingredients.

(ii) Some rules are a matter of choice deemed necessary for the smooth functioning of the society; for example, each country enforces either the left or right-hand side traffic. At the root, the rule is scientific because it is based on the laws of physics, which describe the results of collisions. However, the choice between alternative actions (such as the side of the road on which we have to drive) is arbitrary. In practice, it is based on tradition. Sweden switched from left-hand side to right-hand side traffic in 1967—this shows that the choice of the side is mainly arbitrary and may be dictated by secondary considerations, such as compatibility with other countries.

(iii) Some rules concern the structure of society. They include rules about marriage and inheritance. They are not scientific, but they do have predictable effects on the overall functioning of society.

(iv) Some rules seem arbitrary. They are often strictly enforced because they send social messages and create bonds. For example, people are expected to dress elegantly for weddings and funerals. The dress code has varied greatly over time and still varies significantly across the globe. Fashions in everyday wardrobes, home decor, food, etc., are gently enforced by social pressure.

The first two types of rules are enforced by law, and the same applies to the rules in the third category in many societies.

Members of society who do not follow these rules can be punished, sometimes even severely.

The attitude towards breaking the last set of rules is the most relaxed. People who dress unusually are often tolerated, especially if they are liked or admired for other reasons.

There is a vast number of rules based on probability. These are typically concerned with safety, security or health. The list of examples is very long and includes the maximum driving speed, fastening of seat belts, drinking and driving, codes for home electrical installations, selling alcohol and tobacco to minors, food safety regulations for restaurants, such as a minimum or maximum temperature for food storage, and protective helmets (hard hats) at construction sites. All these rules are probabilistic because they do not result in deterministically predictable consequences. At best, they can reduce the statistics of injuries and economic losses.

As far as I can tell, advertising unusual subjective probabilities is a taboo. A tobacco company cannot say in an ad that smoking lowers the probability of lung cancer. An alcohol producer cannot claim in an ad that drinking before driving lowers the probability of an accident.

According to the subjective philosophy of probability, there is no objective probability. Hence, one cannot enforce probability-based rules for the same reasons as rules in categories (i) and (ii).

When many people have to coordinate actions that may affect all of them, it is reasonable to let the majority decide the rule. Anybody can be injured or killed by a drunken driver. Hence, drinking and driving may be forbidden because of the alignment of subjective probabilities of the majority of people related to drinking and driving.

The same argument sounds much less convincing when it comes to seat belts and cigarette smoking by minors. The potential damage to other people is minimal and cannot, in my opinion, outweigh the restriction of individual freedoms. People should be allowed not to fasten seat belts, and parents should be allowed to let their children buy cigarettes if we adopt the subjectivist stance that there is no objective probability.

I am unaware of a subjectivist discussion of individual freedoms related to probability-based rules enforced by society. In practice, subjectivists seem to accept the same rules as everybody else. In other words, subjectivists accept that probability is objective for all practical purposes despite their claims to the contrary.

14.5.6 *Subjectivist belief in frequencies*

Consider $n + m$ observations of random quantities and denote them as $X_1, X_2, \dots, X_{n+m}$. I will focus on a simple version of the example, in which each random variable X_k may take only values 0 or 1. This is equivalent to observing an event and encoding the result as 1 if the event occurred and 0 otherwise. Think about tossing a deformed coin and let the "event" be heads. Then

$$Y_n = \frac{1}{n}(X_1 + X_2 + \cdots + X_n)$$

and

$$Z_m = \frac{1}{m}(X_{n+1} + X_{n+2} + \cdots + X_{n+m})$$

are relative frequencies of the event in the first n observations and the next m observations, respectively.

Suppose that the sequence $(X_1, X_2, \dots, X_{n+m})$ is exchangeable, that is, if $\pi(1), \pi(2), \dots, \pi(n+m)$ is any permutation of integers $1, 2, \dots, n+m$ then the sequences $(X_1, X_2, \dots, X_{n+m})$ and $\left(X_{\pi(1)}, X_{\pi(2)}, \dots, X_{\pi(n+m)}\right)$ have the same distribution. It follows easily from de Finetti's Theorem (representing exchangeable sequences as mixtures of i.i.d. sequences) and the Law of Large Numbers that for any $p < 1$ and $\varepsilon > 0$ there exists $k < \infty$ such that if $n, m \geq k$ then

$$P\left(|Y_n - Z_m| \leq \varepsilon\right) \geq p. \tag{14.1}$$

This result holds for all "extendable" exchangeable sequences; that is, we must assume that $(X_1, X_2, \dots, X_{n+m})$ can be extended (in principle) to an arbitrarily long exchangeable sequence (see p. 137 of [6]). This is a prevalent assumption in

Bayesian statistics. Most of the time, the term "exchangeable" is used as a synonym of "extendable exchangeable."

An important feature of (14.1) is that k does not depend at all on the "mixing measure," in other words, on the Bayesian prior distribution. So, if X_k's represent tosses of a deformed coin, one can choose k that will work for all prior opinions on how the coin is biased (or not biased). I find it especially striking that the prior opinion could even be "pigheaded" in the sense of [6], p. 114. So even if you "know" beforehand that the probability of heads is less than 1/3 or that it is exactly equal to 0.1372, then (14.1) applies to your probabilistic opinion.

I consider (14.1) a confession of the frequentist faith. This technical result is all that is needed to implement the frequency interpretation of probability in practice. Hence, Bayesians who believe in exchangeability are effectively frequentists. I do not recall (14.1) discussed in the subjectivist literature. If (14.1) is accepted by Bayesians, we can dismiss the philosophical dispute between frequentists and subjectivists as irrelevant from the scientific point of view.

There are many practical examples where we are interested in the relationship between relative frequencies Y_n and Z_m. For example, suppose that a pharmaceutical company developed "su perpenicillin," a new antibiotic, and would like to know the rate of side effects. Then Y_n would represent the relative frequency of side effects in clinical trials, and Z_m would represent the relative frequency of side effects in the population of patients taking the drug after its commercialization, say, over the next 20 years. Exchangeability is a common assumption in most situations involving manufacturing substances and objects. Under this assumption, (14.1) shows that if the numbers of patients in the clinical trials and the general population are sufficiently large, then one should believe that the relative frequencies of side effects in the clinical trials and the general population will be very close with very high probability.

We can turn (14.1) into the following philosophical statement,

$$\text{frequency}(1) = \text{frequency}(2). \qquad (14.2)$$

In other words, one should expect equality of frequencies under appropriate assumptions on the model. The frequentist version of (14.2) is

$$\text{frequency}(1) = \text{probability} = \text{frequency}(2). \qquad (14.3)$$

In other words, a frequentist would say that the probability of side effects is very close to the observed relative frequency Y_n. The frequency Z_m will also be very close to the probability of side effects. This implies that Z_m will be very close to Y_n.

The difference between subjectivist and frequentist views of the world is that frequentists add the middle term in (14.3). It seems perfectly reasonable to claim that the middle term is irrelevant. After all, the patients and the health care system would be affected by Z_m, not by some abstract number called "probability" in (14.3). Given this attitude, it is hard to see the difference between the subjectivist and frequentist views of frequencies. Both communities believe observed frequencies in two appropriately related sequences will be very close, which is all that matters.

Let's analyze a scientific example from a different field to understand why the frequentists identify frequency and probability. Suppose a space agency wants to send a satellite to an orbit around the Moon. They will take the following steps. First, they will observe the trajectory of the Moon. They will use the observations to calculate the gravitational pull of the Moon. This will be used to calculate the parameters of the satellite's flight. The procedure can be represented as

$$\text{trajectory}(1) \Rightarrow \text{gravity} \Rightarrow \text{trajectory}(2). \qquad (14.4)$$

Do we need the middle term, "gravity," in (14.4)? Why not simplify the procedure as follows,

$$\text{trajectory}(1) \Rightarrow \text{trajectory}(2)? \qquad (14.5)$$

After all, the agency cares about the proper placement of the satellite in orbit, not about some abstract "gravity."

The relativity theory shows that one can think about the motion of objects with mass as a motion in a curved space without explicit reference to gravitational force. But it is obvious why "gravity" remains a scientific concept. The ontological status of gravitation is irrelevant. Gravitation is a convenient way to build an image of the universe in the human mind. Very few people have mastered relativity theory to the point that they can abandon gravitation in their understanding of the world. On the technical side, gravitation is an excellent tool in mathematical physics on the human scale and even on the small astronomical scale. So even if gravitation is total fiction, it is beneficial as a tool to compute scientific quantities. My remarks resemble some positivist ideas, but I believe that they are so simple that they can stand on their own without the support of any major epistemological theory.

Returning to the middle term in (14.3), called "probability," it may be true that probability is total fiction. However, this does not matter in practical applications as long as probability calculations yield verifiable and desirable results. Insisting that (14.2) is more rational than (14.3) is like insisting that the relativity theory shows that (14.5) is more rational than (14.4). Even if such a claim is true in some abstract philosophical sense, it is hard to see why it would be helpful in practical life.

The advantage of introducing "fictitious" quantities "probability" in (14.3) and "gravity" in (14.4) is that they can serve as memory in case "frequency(2)" and "trajectory(2)" are not yet known. The values of probability and gravity can be reused in many decision problems in the future. Probability and gravity are convenient ways of compressing and storing the data.

I finish this section with a discussion of a few arcane points so that it is clear that I did not miss them and that I am not trying to hide anything under the rug. First, the example in this section was based on the "mild" interpretation of subjectivism. Recall that according to strict subjectivism, objective probability does not exist. Hence, no consistent distribution is closer to the truth than any other. An opinion about a sequence of random variables that makes them exchangeable is only one particular way to eliminate Dutch books. Thus, this opinion has no relationship to frequencies Y_n and Z_m that will be observed in the future. This is because there are many consistent opinions about the sequence of X_k's, and they put Y_n and Z_m in all kinds of relationships. Since none of these opinions is close to the truth (because there are no true probabilities), there is nothing rational that a subjectivist can say about the relationship between the frequencies Y_n and Z_m. All that de Finetti's theory offers is an algorithm for eliminating Dutch books.

Second, according to the prior opinion, (14.1) asserts the high probability of the approximate equality of the frequencies Y_n and Z_m. In real life, most of the time, the relevant opinion is the one formed after observing $X_1, X_2, \ldots, X_n$ (and only these random variables). Recall the example involving a new drug. The company has to decide on the marketing of the new drug after collecting data from clinical trials, not beforehand. The Bayesian method would predict that (unknown) Z_m will be close to (known) Y_n if the prior distribution were not far away from the uniform distribution, that is, if the prior opinion about the probability of side effects were evenly spread, more or less, over $[0, 1]$. However, the prior does not have to be uniform. Now I have to split my analysis into two cases.

There is nothing else to say when the opinion holder is a strict subjectivist. See my previous discussion. All consistent opinions are equally valid, so those making Z_m and Y_n distant are as valid as those that make these frequencies close.

If the prior distribution is related to the real world and is not spread equally, more or less, over $[0, 1]$, this means that there is a reason for the departure from uniformity. In other words, the opinion holder has extra information that may take various forms. In the new drug example context, this may be knowledge of similar drugs and their side effects. In this case, it is perfectly rational to maintain that Z_m does not have to be close to Y_n. However, a similar conclusion applies to the frequentist analysis. Extra information should not be ignored. A rational frequentist approach is to build a model that can accommodate the extra available information. Then Z_m and Y_n need not be close.

14.5.6.1 *Measuring subjectivist commitment*

I propose a practical procedure to measure the strength of subjectivist beliefs. Recall the example and notation at the beginning of Sec. 14.5.6. Suppose that Barbara is a subjectivist. She will be given a choice between two decisions. I will phrase the example in terms of loss rather than gain to match the customary measure of statistical inaccuracy.

Barbara will not be told what physical device generates the sequence $X_1, \ldots, X_{n+m}$. She will be assured that the sequence is exchangeable.

Barbara will be given the values of $X_1, X_2, \ldots, X_n$ and then asked to guess the value of Z_m. If her guess is p then she will lose $(p - Z_m)^2$ dollars. This represents the "square loss," widely used in theoretical statistics.

Barbara will be asked to choose between two possible statistical procedures. The first procedure D_1 is to take $p = Y_n$. In other words, the first choice is to apply (14.3), the formula embodying the frequentist belief that probability and frequency should be identified.

The second option D_2 is to use the Bayesian approach. Barbara could assume that the sequence $X_1, \ldots, X_{n+m}$ is exchangeable since she was assured this was the case. If she does, she will choose a prior distribution and then calculate the conditional (posterior) distribution of Z_m given Y_n. Barbara can choose any functional of the posterior distribution as her guess p, whether she uses the exchangeable model or not. But there is a catch. Since Barbara is a subjectivist and D_2 is the Bayesian approach, D_2 is supposedly better than D_1. If Barbara chooses D_2, she must pay extra w dollars.

What value of w would make D_2 more attractive than D_1 to a subjectivist? The answer depends on n and m, so consider a specific example: $n = 30$ and $m = 10,000$. I have chosen $n = 30$

to represent a small sample size in the frequency statistics and $m = 10,000$ to represent a sample that should give an accurate estimate of the "true" probability according to supporters of the frequency interpretation of probability.

The critical value of w that makes D_2 better than D_1 depends, of course, on the decision-maker. I guess the critical value of w would be very small for most Bayesian statisticians. If I am right, the subjectivist refusal to identify frequencies and probabilities is an ideological position with little or no practical meaning in elementary statistics.

14.5.7 *People are animals*

I believe that animals use probability, and their use of probability is an excellent illustration of and strong empirical support for my laws (L1)-(L6). But this is not my main point. As elsewhere in this chapter, the main point is that the rather obvious problem of probabilistic behavior among animals is not a part of the mainstream philosophical discussions of probability (to my best knowledge).

We do not know what animals think. I doubt that we will understand how they perceive the world, make decisions, etc., any time soon. Nevertheless, we can adopt the behaviorist approach and guess how animals make decisions based on observations of their behavior. Animals have to make decisions in situations involving randomness, for example, when they choose where to search for food, how to catch prey, how to avoid being preyed upon, how to fight a predator, where to build a nest, etc. The theory of evolution implies that individual animals that make good decisions are more likely to have descendants than animals that make bad decisions. The genes responsible for high success rates will be passed to the offspring. Hence, good survival strategies are reinforced in the animal kingdom.

Everyday observations of domesticated animals, observations of wild animals, and laboratory experiments show that animals can learn. More specifically, animals can detect frequencies in the sense that they choose those actions that were rewarded most often in the past iterations of a real-life situation or an experiment. A simple explanation of this behavior is that an animal identifies the observed frequency with probability and then uses this probability to make the optimal decision. Of course, my argument is anthropomorphism. We do not expect animals to be able to grasp the concepts of frequency and probability consciously. Still, identifying frequency and probability allows us, humans, to understand animal behavior.

A possible philosophical stand is to abandon attempts at any higher-level interpretation of animal behavior. One can record the pattern—animals choose actions that were rewarded with the highest frequency in the past. However, this opens the question of whether this interpretation should be applied to humans. After all, humans are animals, and it seems obvious that human evolution should have created many instinctive human behaviors similar to those of instinctive animal behaviors. Suppose we acknowledge that animals are successful (have high "fitness") because evolution created the instinct to repeat actions that were most beneficial in the past. Why shouldn't we apply the same strategy? We can observe and formally record frequencies. Our "frequency instinct" tells us to take actions rewarded most often in the past.

In my interpretation of animal behavior, frequency is first identified with an abstract probability value, which is then used in a single decision problem. The decision problem may be recurrent but does not have to be. My interpretation does not agree with von Mises' theory because animals apply probability to single events. As for the subjective philosophy of probability, I do not see how the Dutch book argument could explain animal behavior. While I cannot prove that animal behavior is equivalent to human decision making in the face of uncertainty or that it provides definitive proof of my theory (L1)-(L6), the lack of significant discussion of animal behavior in the major presentations of the subjective and frequency philosophies of probability leaves a gaping void.

14.5.8 *Empirical evidence of objective probability*

Nothing can be proved in philosophy. But why not try?

There is an unlimited amount of empirical evidence that objective probability does exist. The whole scientific literature is filled with such evidence. I do not have enough space to copy the whole scientific literature here, so I will present just one example, taken from [83]. Weather prediction methods have greatly progressed in recent decades. If one graphs the frequency of rain against the probability of rain given in the forecast, the resulting curve is almost a straight line from $(0, 0)$ to $(1, 1)$. In other words, the probabilistic rain predictions are almost perfect. Doesn't it prove the existence of objective probability and scientists' ability to estimate it reliably? Subjectivists may disagree. But then they must explain how the Dutch book argument explains the excellent match between weather predictions and observed frequencies.

The standard subjectivist argument against the frequency interpretation of probability goes like this

$$\text{frequency} \not\Rightarrow \text{probability}. \tag{14.6}$$

Yes, even the mathematical theory of probability does not say that the observed frequency of an event in a finite sequence of observations necessarily equals the probability of the event (if objective probability exists). But the weather prediction example can be represented as

$$\text{probability} \Rightarrow \text{frequency}. \tag{14.7}$$

If an event occurred in the past, many theories can be proposed to explain it, and this is typically done. Consider multiple historical explanations of wars and revolutions or why the stock market increased or decreased in a given year. Now, consider predictions of market performance in the years to come. If anyone gets these right, she will be admired. There have been examples of excellent market predictions made in the past, but often, they were followed by less impressive performance.

The scheme (14.6) looks very much like an "explanation of past events," and hence, it is not very impressive or convincing.

However, weather predictions encapsulated in (14.7) represent a great practical success. This piece of empirical evidence cannot be summarily dismissed. It deserves careful analysis from subjectivists.

Since we are discussing empirical evidence, let us consider the following artificial but doable experiment: The experimenter will generate a sequence of one hundred 0s and 1s. Two participants will be offered the same bet on the experiment's outcome. The first participant will not be told how the sequence will be generated. The second participant will learn the physical mechanism for generating the data. For example, the first five bits (0's or 1's) will be generated by the tosses of a fair coin. The tosses of a deformed coin will generate the following five bits. The next ten bits will be generated by rolls of a die, with 1 representing four dots. The following 20 bits will be generated by sampling cards from a deck of 52 cards, with replacement, with 1 representing the queen of hearts. And so on. Both participants can be consistent so that no Dutch book will be created, assuming they choose to be consistent. It is evident that, depending on the bets, the second participant can win much more money than the first. I do not know how subjectivists would react to this simple thought experiment. The examples I found in the subjectivist literature do not provide any clue (to me).

14.5.9 *Paranoid decision-makers*

De Finetti's idea that we should use probability to protect ourselves against Dutch books is effectively a case of paranoia. According to [84], "Paranoia is the feeling that you're being threatened in some way, such as people watching you or acting against you, even though there's no proof that it's true."

Engineers designing an airplane make computations for one particular design or several competing designs but not all possible airplane designs. They do only as many mathematical calculations as needed for the purpose. This saves time, labor, money, etc. The idea that Bayesians should make extensive calculations to protect themselves against all possible Dutch books is totally impractical. If Bayesians wanted to protect themselves against Dutch books, they would find it much easier to review all the relevant potential contracts or decisions (typically, only a small number of these are available at a given time) and see whether they form a Dutch book or not.

14.5.9.1 *Can you trust anyone?*

Consider a simple experiment testing human attitudes toward decision making. The participants will be informed that they can make one specific bet on the result of 100 tosses of a deformed coin. They will be told explicitly that there will be no more bets on the outcome of the tosses. If the bet pays some money for some outcomes, should a Bayesian participant use probability to eliminate all possible Dutch books potentially created by bets unavailable during the experiment? Should one trust the experimenter when she says there will be only one available bet, or must you protect yourself against all possible Dutch books? According to de Finetti's theory, probability is unnecessary if it is rational to trust the experimenter.

14.5.9.2 *Tabula rasa*

Consider a paranoid decision maker who wants to protect himself against all possible Dutch books. De Finetti has good advice for him: use probability to coordinate your decisions.

Now suppose that the decision maker obtained some new data. According to the standard Bayesian procedure, he will replace his prior distribution with a posterior distribution. At this point, he knows that he will never use his prior distribution again. From now on, he will use his new posterior distribution and its successors exclusively. Suppose he searched for all decisions he made using the prior distribution and determined that there were none. No matter how paranoid the decision maker is, he is now sure he does not need to coordinate any prior decisions with the posterior decisions. It follows that he can take any probability distribution as his posterior distribution.

The situation outlined above is common in scientific research. Typically, no decisions are based on the prior distribution. This distribution usually plays a strictly technical role in statistical analysis and is not the basis for any decisions. De Finetti's theory emaciates the Bayes theorem.

14.5.9.3 *No need to use probability*

In Sec. 6.2, I have shown that the Dutch book argument is useless. Still, it may be interesting to see how the Dutch book argument fails in a simple toy model representing many applied decision problems. We will assume that the decision-maker is not paranoid. She believes the family of decisions available to her is as stated in the model.

Suppose that a future event A is random so that it may occur or not. Next, consider an event B representing "information." To make things interesting, let us assume that all four events (A and B), (A and not B), (not A and B), and (not A and not B) can occur. We assume that the outcome of B is observed before the outcome of A. Another simplifying assumption is that there are only two decision problems.

First, one has to choose between two options, O_1 and O_2, before observing B. The payoffs depend on the outcome of A, but they do not depend on B. There is a p between 0 and 1 such O_1 is preferable to O_2 (that is, the expected gain is greater in the case when one makes decision O_1 than in the case O_2) if $P(A) < p$. If $P(A) > p$ then O_2 is preferable to O_1. Finally, O_1 and O_2 are equally attractive if $P(A) = p$.

The second decision problem is similar to the first one except that a decision has to be made after observing B but before observing A. Hence, the relevant probability is $P(A \mid B)$, not $P(A)$. One has to choose between two options, O_3 and O_4. The payoffs depend only on the outcome of A. There is a q between 0 and 1 such O_3 is preferable to O_4 if $P(A \mid B) < q$, O_4 is preferable to O_3 if $P(A \mid B) > q$, and O_3 and O_4 are equally attractive if $P(A \mid B) = q$.

It is elementary to see that if $0 < p < 1$ and $0 < q < 1$, then there exists a probability distribution on the space carrying events A and B such that $P(A) < p$ and $P(A \mid B) < q$. If a decision-maker adopts this probability distribution, she will take actions O_1 and O_3. Similarly, there exists a probability distribution such that $P(A) > p$ and $P(A \mid B) < q$. If a decision-maker adopts this probability distribution, she will take actions O_2 and O_3. The same argument applies to the remaining two possibilities. We conclude that each one of the four possible decision strategies (O_1, O_3), (O_2, O_3), (O_1, O_4), and (O_2, O_4), can be adopted by a subjectivist decision-maker.

It follows that all four decision strategies (O_1, O_3), (O_2, O_3), (O_1, O_4), and (O_2, O_4) are objectively rational. According to the subjective philosophy, since all four possible decision strategies are objectively rational, there is no need to apply probability. One can choose one of the four strategies in a completely arbitrary way. We conclude that probability is useless in this context.

14.5.9.4 *Applications*

Although highly simplified, I claim that the model presented in the last section represents the essence of much of applied probability. The assumption that the probability space is divided into only four events (A and B), (A and not B), (not A and B), and (not A and not B) is not essential. What matters is that the payoffs of the actions O_1, O_2, O_3 and O_4 are based only on the event A and not at all on B.

Applied statistics can be divided roughly into two cases: when data have their own value, irrespective of any value they have as information, and when the opposite is true. To illustrate the first case, note that when we speak of "collecting statistics" outside the scientific context, we often mean information on births, deaths, wars, trade, unemployment, college graduations, divorces, cancer, car accidents, marriages, etc., that is, information on the functioning of the society. In all these examples, individual events are likely to have significant (positive or negative) utility to some individual people. They also have "global" utility to communities, countries, social classes, etc. In this case, one should account for the event B in the utility function.

Scientific experiments best represent the other case. The results of experiments often have no direct value, their value being exclusively derived from their contribution to the information pool. Consider a chemical manufacturer working on a new fertilizer. The outcome of experiments with some plants given the new fertilizer has no value except for the information.

I should point out that experiments cost money, but this is a non-random upfront cost; therefore, it does not enter the utility function in a way that would be relevant to the probabilistic analysis.

The example discussed in Sec. 14.5.9.3 encapsulates the essence of much of applied statistics, specifically, statistics concerned with analyzing results of experiments, astronomical observations, archaeological discoveries, etc. In all these cases, de Finetti's theory implies that the Bayes theorem is useless because prior and posterior decisions cannot conflict.

14.6 Frequency philosophy of probability

I will present examples where the frequency theory fails to provide a foundation for established scientific methods. In this section, I will not try to distinguish between collectives and i.i.d. or exchangeable sequences because I will not be concerned with the differences between these concepts. Instead, I will discuss their common limitations.

14.6.1 *Single events with no probability*

The essence of von Mises' philosophical theory is that probability does not apply to single events. Von Mises said that ([3], p. 11),

> We can say nothing about the probability of death of an individual [within a year] even if we know his condition of life and health in detail.

I am missing examples illustrating the following two questions.

If probability does not apply to single events, can one give sensible advice to decision-makers dealing with such cases?

Which events should be considered individual, outside any collective?

Both questions need clarification. They may appear to be independent but are closely intertwined.

Let us consider the following example, an elaboration of von Mises' own example. Suppose that two patients arrive at an emergency room (ER) simultaneously. Only one ER doctor is available, and she can treat only one patient at a time. The first patient has chest pain and other symptoms of a heart attack. The second patient cut his finger. The bleeding stopped, but the cut was so deep that it needs stitches. Suppose that the minor surgery that the second patient needs would take 30 minutes, and the first patient could die within this time. I believe that the doctor would treat the patient with a possible heart attack first.

Von Mises' statement may be interpreted in two radically different ways. According to von Mises, there is no way to tell which of the two patients has a higher probability of dying in the next 30 minutes (more precisely—these probabilities do not exist). Hence, it is rational and equally ethical for the doctor to treat any of the two patients first.

According to the second interpretation, we should consider all patients with chest pain and cut wounds. For each of the collectives, we should consider long-run frequencies of death within 30 minutes of arrival at an emergency room, in those cases when they are treated and, separately, in those cases when they could not be treated within this time. The analysis of these observed frequencies would show that, in the long run, treating patients with chest pain before patients with cut wounds would save more lives than if they had been treated in the opposite order.

I guess that most frequentists would go with the second interpretation. But then, wouldn't it be simpler to say that the patient with chest pain has a higher probability of dying in 30 minutes than a patient with a small cut wound?

The analysis of the above example points to a problem with von Mises' theory. The theory may be vacuous because every event belongs to a collective, and the decision problem based on a single event can always be interpreted as an element of a collective of decision problems.

The following is an example of an isolated event that does not fit naturally into any collective. Consider global warming, nowadays referred to as climate change. Global warming will occur or will not occur in the next 50 years. What collective could this event possibly fit into? One could argue that although there is no such collective, any economic decisions made by governments about possible global warming are elements of the collective of all economic decisions. This proposal stretches the definition of a collective because economic decisions are hardly exchangeable or i.i.d. in the mathematical language. Hence, they do not form a collective in von Mises' sense. However, my point remains that frequentists do not supply examples of events with probabilities assigned by representatives of other philosophical directions that would not be given any probabilities according to the frequency interpretation. Moreover, there are no examples illustrating differences in decision making. How would decisions made by frequentists in single-event cases differ from those made by people who believe in probabilities of single events?

14.6.2 *Stochastic processes*

A large number of sequences of random variables encountered in scientific practice and real-life applications are not i.i.d. or exchangeable—it is a tradition to call them "stochastic processes." Some of the best-known classes of stochastic processes are Markov processes, stationary processes and Gaussian processes. Markov processes represent randomly evolving systems with short or no memory. Stationary processes are invariant under time shifts; if we start observing the process today, the observations will have the same probabilistic characteristics as if we started observations yesterday. Gaussian processes are harder to explain because their definition is somewhat technical. They are closely related to the Gaussian (normal) distribution, which arises in the Central Limit Theorem and has the characteristic bell shape. One can make excellent predictions based on a single trajectory of any of these processes. Predictions may be based on various mathematical results such as the "ergodic theorem" or the extreme value theory. In some cases, one can transform a stochastic process mathematically into a sequence of i.i.d. random variables. However, even when this is possible, this purely mathematical procedure is artificial and has little to do with von Mises' collectives. The frequency philosophy is useless when we try to apply it to stochastic processes because the predictions mentioned above do not correspond to frequencies within any real collectives.

As a more concrete example, consider two casinos that operate in two different markets. Suppose that the amount of money gamblers leave in the first casino can be reasonably modeled as an i.i.d. sequence. In the case of the second casino, assume that there are daily cycles because the types of gamblers that visit the casino at different times of the day are different. Hence, the income process for the second casino cannot be modeled by an i.i.d. process. Suppose that in both cases, we were able to confirm the models and estimate the parameters using data from the past. In each case, we can make a prediction for earnings of the casino over the next year. The theory of von Mises can be applied directly to make an income prediction in the case of the first casino. In the case of the second casino, the frequency theory says that we have to find an i.i.d. sequence (collective), presumably a long sequence of similar casinos, to apply the probability theory. This is totally unrealistic. In practice, predictions for both casinos would be considered equally valuable, whatever that value might be. Nobody would even think of finding a sequence of casinos in the second case.

14.6.3 *Unlikely events*

Another class of examples where the frequency theory is miles apart from real science are situations involving very small probabilities. Suppose someone invites you to play the following game. He writes an arbitrarily chosen 20-digit integer on paper without showing it to you. You have to pay him \$10 for an opportunity to win \$1,000, if you guess the number. Anyone who has even a basic knowledge of probability would decline to play the game because the probability of winning is a meager $9 \cdot 10^{-19}$. According to the frequency theory, we cannot talk about the probability of winning as long as there is no long-run of identical games. The frequency philosophy has no advice to offer here, although no scientist would have a problem with rational choice. Practical examples involve all kinds of very unlikely events, for example, natural disasters. Some dams are built in the US to withstand floods that may occur once every 500 hundred years. We would have to wait many thousands of years to observe a sequence of floods sufficiently long so that the observed frequency could be matched, more or less, with the theoretical probability. In that time, several new civilizations might succeed ours. According to the frequency theory, discussing the probability that a dam will withstand floods for the next 100 years makes no sense. Even more convincing examples arise in the context of changing technology. Suppose that scientists determine the probability that there will be a serious accident at a nuclear power plant in the US in the next 100 years is 1%. Suppose the estimate is correct. In that case, one must wait about 10,000 years to observe one or a handful of accidents or wait for 100,000 years to get a solid statistical confirmation of the probability value. This is totally unrealistic because the technology is likely to change drastically much sooner than that, say, in 100 years. I do not think anyone can imagine what nuclear power plant technology will be 1,000 years from now. The 1% estimate has a lot of practical significance, but it cannot be related to any "long-run" of observations that could be made in reality.

There are many events that are not proven to be impossible but have probability so small that they are considered impossible in practice, and they do not fit into any reasonable long-run of events. It is generally believed that Yeti does not exist and that there is no life on Venus with very high probability. If we take the frequency theory seriously, we cannot make any assertions about probabilities of these and many other equally unlikely events—this is a sure recipe for the total paralysis of life as we know it

14.6.4 *Graphical communication*

It is common in science to present the results of research using graphs such as in Figures 14.2–14.4. Figures of this kind often represent simulations. In a typical case, the figure is supposed to make a visual statement that would be hard to convey by words. The figure usually shows a "typical" shape of a random function or some other random mathematical object. Figures 14.2–14.4 show simulations of fractional Brownian motion with three different values of the "Hurst parameter." The graph becomes smoother as the parameter becomes larger. In each of the three cases, one of the possible shapes of the fractional Brownian motion is a straight line, depicted in Figure 14.5. Strictly speaking, the trajectory cannot be perfectly straight but can be straight with any given accuracy. In particular, it can appear straight to humans, with strictly positive probability. The probability of seeing a straight trajectory of fractional Brownian motion with a parameter less than, say, 0.95 is extremely small. The message conveyed by Figures 14.2–14.5 is that fractional Brownian motion processes take the shape of a straight line with negligible probability. Moreover, the paths of fractional Brownian motions with different values of the Hurst parameter differ in roughness with overwhelming probability. We do not have to (although we can) look at a long sequence of such simulations to "get the picture." One look at the graphs in Figures 14.2 and 14.4 is all that one needs to tell which of the two processes has a larger Hurst parameter.

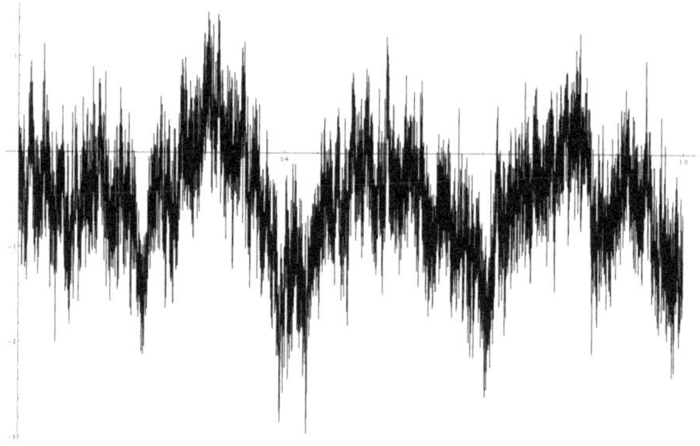

Fig. 14.2 Fractional Brownian motion with Hurst parameter 0.1.

For readers who have some technical knowledge of stochastic processes I would like to point out that I used fractional Brownian motion as my illustration because its increments are not i.i.d. Suitable transformations can be used to represent the process as a function of an i.i.d. sequence but this has nothing to do with von Mises' collectives.

Snapshots of complex stochastic models are an excellent way to communicate claims about small probabilities. They do not fit into the frequency interpretation of probability in any reasonable way. The absence of their discussion in the philosophical literature illustrates my point about how missing examples allow dubious philosophical theories to escape scrutiny.

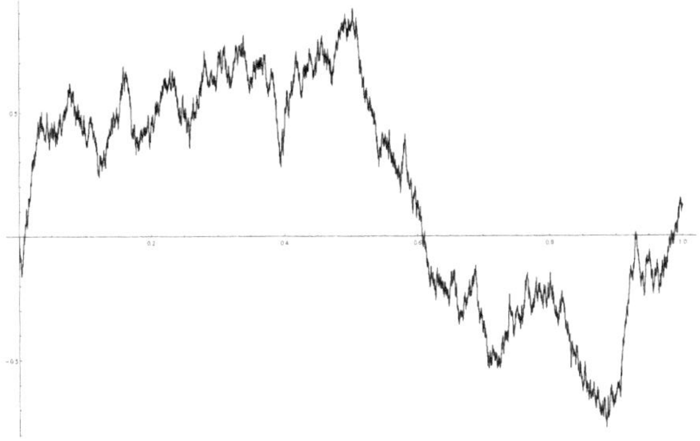

Fig. 14.3 Fractional Brownian motion with Hurst parameter 0.5 (standard Brownian motion).

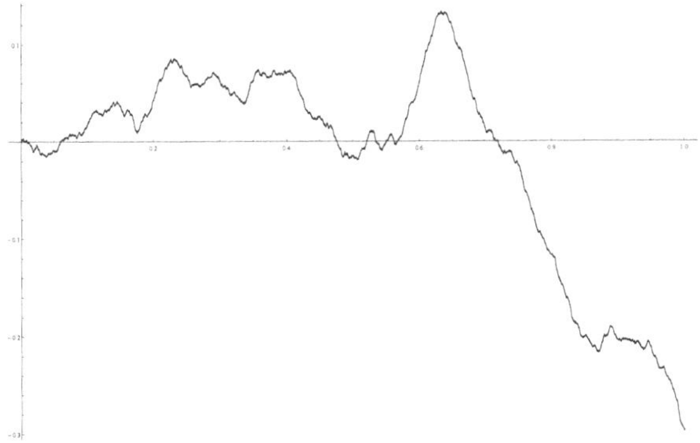

Fig. 14.4 Fractional Brownian motion with Hurst parameter 0.9.

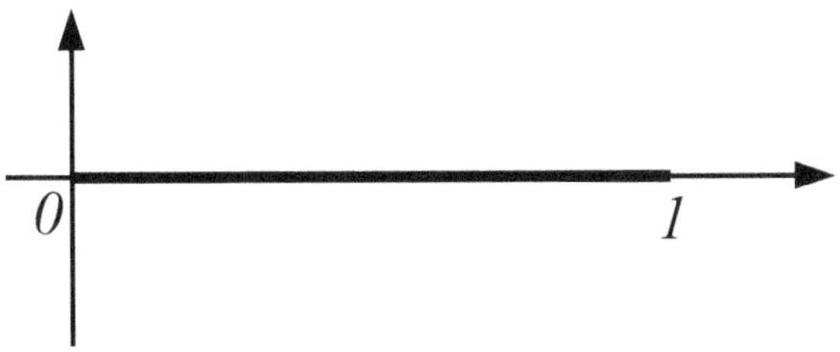

Fig. 14.5 The straight line in this figure is a possible trajectory of a discrete approximation to fractional Brownian motion with any Hurst parameter. The probability of such a trajectory is strictly positive but extremely small.

Chapter 15

Insular Ideological Communities

The subjective philosophy of probability has common defensive mechanisms with diverse ideologies. I will identify and analyze some of these mechanisms.

Why are some ideologies highly popular within some communities but entirely rejected by other groups? I will argue that there are many "Insular Ideological Communities" (IIC) with their own standards of excellence, effectively erecting insurmountable barriers between their supporters and their critics.

The practical needs of an IIC, such as making money or meeting spiritual needs, shape the theory and practical interpretation of the ideology. This observation is analogous to the Marxist claim that the practical needs of the ruling class determine the ruling ideology.

My interest in IICs stems from my disappointment with the philosophy of probability, especially outside the professional philosophical community, on the outskirts of science. For a considerably different general view of ideologies from the psychological perspective, see [85].

15.1 Intellectual standards

Scientists may believe that their method of analysis of the natural world is the best of all. However, the existence of various non-scientific ideologies and intellectual endeavors strongly suggests that human needs are so complex that alternative approaches are as attractive as science to many (perhaps all) people, at least in some circumstances.

15.1.1 *Science*

I have presented my view of the foundations of science in Chap. 8. The central aspect of science relevant to the present discussion is its universal character in human society. Philosophers proposed many definitions of science. Here is another one. Science is the collection of facts, theories and methods everyone understands and accepts, regardless of nationality, ethnicity, religion, culture, etc.

The definition given above is oversimplified. Some of the exceptions to the rule are not very interesting. For example, some people with mental problems, such as paranoia, may reject facts established scientifically. A more interesting example is provided by "creationism." According to [86], "[...] 2017 Gallup creationism survey found that 38% of adults in the United States inclined to the view that 'God created humans in their present form at one time within the last 10,000 years' [...]." Creationists reject Darwin's evolution theory, which is widely considered scientific. Creationists do not reject the evolution theory because they find it arbitrary or unproven. They derive their opinions from a holy book and try to find cracks in the evolution theory. In this way, they implicitly recognize the universal value of scientific reasoning.

The strength of science is its applicability in everyday life. This includes simple science, such as arithmetic, used by shoppers and sellers. The applicability includes the most advanced science, such as quantum theory, because many currently widespread devices, such as cell phones, include chips based on quantum physics.

Travelers know they can count on the worldwide universal understanding of many concepts affecting ordinary life, including numbers and the associated arithmetic, water, food, poison, heat, temperature, rain, day, night, weight and electricity.

All of these are in the domain of science. Non-science starts when the concepts and their practical implementations become non-universal. Examples include politeness, marriage, honor, beauty, justice, equality, and good and evil.

Science includes scientific research. I do not identify this activity with science. Scientific research methods are borrowed and adapted from everyday life, but evaluating their results is subjective. Applicability of scientific theories and individual results is highly valued, but it is not the only criterion applied in the assessment of scientific research. Purely subjective or somewhat subjective criteria strongly influence the overall direction of scientific research. These criteria include significance, elegance, generality and simplicity of theories. The result is that the scientific community is divided into sub-communities. Some focus on the most significant and promising directions in contemporary science. Many small and diverse communities work on totally insignificant questions. They strictly adhere to the formal rules of scientific research, such as logic and the induction principle. Within these communities, loyalty to the formal rules is considered a sufficient condition for admission to science.

15.1.2 *Mathematics*

Strictly formalized writing characterizes mathematics. All articles and books aspiring to belong to mathematics must be rigorous. Mathematicians usually present their results informally, but this is only to make their publications easily accessible to humans. Every definition, theorem and proof must be stated in a way that can be unambiguously translated into formal logic. Although mathematical methodology may seem considerably different from science, the overall functioning of mathematics is very similar to science's. Parts of mathematics are directly applicable, just like parts of science are. The significance of other mathematical contributions is evaluated in a greatly subjective way based on elegance, generality, simplicity, and other rather vague criteria. This allowed some small, tightly-knit communities to arise. Their members support each other even if the general mathematical community ignores them.

Mathematicians create some nonsensical theories for the same reasons philosophers do. Neither mathematical nor philosophical theories have to be applied, so contributors to both fields feel free to produce material that is formally correct and agrees with the general practices in the field but might not have any applications. This can be justified as pure intellectual pleasure or pastime. However, these practices are valuable—they constitute the practice of intellectual skills that can be applied in a more significant context or passed on to future generations.

15.1.2.1 *Mathematical physics*

Mathematical physics is an interesting intellectual direction situated between mathematics and physics. At least some literature in this area is neither rigorous in the mathematical sense nor empirically supported in the scientific sense. The main quality criterion seems to be the self-consistency (in the logical sense) of this area of research.

15.1.3 *Religion*

Religion is a very complex phenomenon. At the level intended for the mass consumption of rank-and-file followers, religion's intellectual content is subservient to its dogmas. At the same time, the top intellectuals enjoy considerable freedom because their thoughts are often very abstract and not likely to be understood by ordinary religious people. There is little danger of the corruption of simple minds by arcane and paradoxical theological theories.

Main religions developed their own concepts, and contributors to the corresponding theologies are judged, among other ways, on their ability to manipulate these concepts.

An interesting aspect of Catholic theology is that it sometimes uses similar methods to science. The approach of the

Catholic Church to miracles illustrates my point. Before a miracle is recognized, the Church collects the data, for example, medical records of an allegedly miraculously cured person, testimony from witnesses, etc. Then, the materials are analyzed, and alternative explanations for the empirical evidence are considered, including standard scientific explanations. Only when these are excluded does the Vatican pronounce a miracle.

15.1.4 *Art*

Art provides an excellent example of an intellectual endeavor that can function very well despite the lack of any formal constraints. Until the second half of the nineteenth century, high art was distinguished by technical standards, aesthetic expectations and human relevance. The technical standards included basic drawing skills (analogous to those in sculpture, writing, etc.), composition, color palette, etc. The human relevance could consist of a political message, a depiction of feelings such as love, a likeness of a person, or a recording of an event for the posteriority.

For various reasons, a revolution in art occurred in the twentieth century. One of its significant aspects is the continuous abandonment of traditional standards. In the following, I will take liberties with the actual chronology. First, photographic realism was abandoned in favor of "impressions" as in Impressionism. Next, the form was grossly distorted in Cubism. Abstraction came next, removing any figurative elements from the paintings. This was followed by Minimalism, often limited to large, simple monochromatic elements. The ultimate piece of modern art was Duchamp's "Fountain," a ready-made urinal displayed at an exhibition in 1917, lacking any connection with the traditional idea of art (see [73]). *Fountain* was not made by the artist, had no particular aesthetic appeal and had no meaning except for the obvious (scatological, offensive). Its main and only function as a piece of art was to offend the viewers and make them think about the essence and meaning of art.

Much modern art escapes from the audience by creating pieces and trends that defy all previously, even recently, established standards. Art lovers have to educate and reeducate themselves continually. Some artists' success seems to result more from their public relations efforts, such as bizarre behavior, than their artistic contributions.

Answer to the quiz posted in Sec. 14.3: All four images are digitally generated (scaled) copies of paintings considered fine art, that is, items displayed in museums or sold by high-end galleries. The top left painting is "5-Bromogramine, 2005" by Damien Hirst. The top right painting is "Two Panels: Green Orange" by Ellsworth Kelly. The bottom left painting is "Composition en rouge, bleu et blanc II" by Piet Mondrian. The bottom right painting is "IKB 79" by Yves Klein.

15.1.5 *Philosophy*

Philosophy has a striking intellectual standard—there are no standards or rules. Philosophers argue a lot, mainly in a civil way but sometimes in an aggressive manner. Nevertheless, no philosophical idea is rejected in the long run, no matter how bizarre. At worst, it may be ignored or forgotten.

I will illustrate my position with an outrageous claim—Alan Sokal's hoax (see [87]) was a genuine contribution to philosophy. Sokal published an article, "Transgressing the Boundaries: Toward a Transformative Hermeneutics of Quantum Gravity" in *Social Text*, a leading journal of postmodern cultural studies, in 1996. The article contained intentional nonsense consisting of a concoction of fashionable ideas and slogans drawn from philosophy and science. The hoax was revealed a few weeks after its publication. The affair attracted much attention. However, the fact that the hoax was not discovered by the editors and readers of *Social Text* proves that Sokal's article was actually a valuable addition to the philosophical literature. Sokal only proved that the quality criteria adopted by postmodern philosophers are utterly different from those adopted elsewhere, for example, in science. Philosophy is an IIC, so it can choose its own intellectual standards. If philosophers think an article meets their standards, it is the real McCoy.

Philosophy is diverse. Relativism, Marxism and the philosophy of probability are all IICs, but each is unique.

15.1.5.1 *Marxism*

Marxism is a philosophical theory with its own characteristics, so it deserves a separate honorable mention. The political side of Marxism was (and is) communism. Many intellectuals supported communism despite extensive empirical evidence that the system was and had to be totalitarian. No empirical evidence of poor economic performance was accepted. The communist/Marxist community was closed to outside criticism. Intellectual opponents were routinely accused of ulterior motives. Marxism had the ambition to be a scientific theory, supposedly empirically justified. Indeed, it made some empirically testable predictions, such as about future revolutions.

15.1.5.2 *Philosophy of probability*

Philosophy of probability is the main topic of this book. This branch of philosophy has some characteristics of science, partly because some of the top contributors (for example, de Finetti and von Mises) were scientists and had scientific instincts, partly because it tries to reach the scientific audience. All primary directions in the philosophy of probability take for granted implicitly the same philosophical assumptions that are tacitly embedded in science: there exists a physical universe, objective truth, and stable laws governing the universe; humans can effectively apply the induction principle; humans can reliably communicate.

15.1.6 *Conspiracy theories*

Conspiracy theories are similar to religion in some respects. Both types of IICs are based on empirical claims not accepted outside the given IIC and address the emotional and social needs of IIC members. However, there are some differences between religion and conspiracy theories. Most religions are complex ideologies addressing various issues, from genesis to ethics, while a typical conspiracy theory is limited to very narrow claims. Many religions are well organized, unlike a typical conspiracy theory limited to events or phenomena in our physical universe.

15.1.6.1 *Anti-vaccination movement*

The anti-vaccination movement is an exceptional conspiracy theory because it is based on scientific claims (see [88]). These claims are not recognized by mainstream science, but the mere fact of referring to science puts the anti-vaccination movement in a league of its own. There are some other conspiracy theories based on empirical claims concerning facts, for example, alleged attempts by the US government to hide remnants of UFOs and alien bodies.

15.2 Rejected transplants

There are different areas of intellectual activity, for example, mathematics, science, philosophy, and theology. They have separate subjects of interest, methods, ideas, and standards of excellence. They partly overlap and partly interact. However, some elements in each of them make no sense if they are transplanted into another field of intellectual activity. Consider the following examples:

(i) The Catholic dogma that there is one God and Holy Trinity makes no sense in mathematics (one is not equal to three).
(ii) Physicists differentiate non-differentiable functions, integrate non-integrable functions, operate on divergent series,

differentiate with respect to integer arguments. None of these operations makes sense in mathematics.

(iii) Mathematicians have a whole industry of proving the existence and uniqueness of solutions to differential equations. Much of this makes no sense to physicists who know that certain equations have solutions.

(iv) Various non-existence claims in philosophy (non-existence of motion, physical universe, objective truth, objective probability) make no sense in science and everyday life.

A field of mathematics gains respect if it can be used to prove results in other fields of mathematics. For example, probability gained respect when it was used to prove non-probabilistic results (see [89]).

A field of intellectual activity gains respect when it contributes to other fields of intellectual activity. Mathematics and physics (more generally, science) contribute to each other. Science contributes to philosophy. Philosophy and theology contribute to each other.

Contributions of philosophy to science and mathematics are hard to identify. Perhaps the obsession of physicists with constructing the fundamental theory of everything is inspired by the theological idea of God as the primary mover.

The philosophical theories of de Finetti and von Mises are significant contributions to philosophy. The two philosophers introduced new concepts—the Dutch book and collective. Transplanting these theories to science is like transplanting the claim that there is one God and the Holy Trinity to mathematics. The result is pure nonsense.

Paradoxically, the subjective philosophy of probability indirectly contributed to science by propping up Bayesian statistics, although this philosophical theory has absolutely nothing in common with this field of science.

15.3 Defensive mechanisms

Some or all outsiders might not share an IIC's intellectual standards. Every IIC can be critiqued based on principles taken from outside its domain. It is perfectly rational for IIC to reject such a critique. Otherwise, no IIC could survive in the long run. Without an absolute arbiter, no principle can survive an attack from outside. This is due to the structure of our logic, which can only deduce new truths from assumed but unproven axioms. IICs who happen to have power often suppress any critique that comes from outside their ranks. Every challenge is considered a mortal danger; outside critics are branded enemies or dilettantes. I list several specific defensive mechanisms in the following sections.

15.3.1 *Coterie mentality*

IICs are often governed, formally or informally, by small coteries of people who hold power to set and interpret the principles of the IIC. Glued together by mutual interest, often economic, the coterie dictates the rules and leads the IIC rank-and-file members.

15.3.2 *Institutionalization*

Some IICs are organized. Institutions such as the Catholic Church, communist parties, and museums often have their own agendas—they partly alienated themselves from the underlying ideology. People assume that institutions provide high-quality theory or practical implementations of the ideology because of their structure, hierarchy, promotion standards, etc. The quality can be distorted by the desire of an IIC's institution to protect its interests. It is natural for an institution to defend its IIC using political means.

15.3.3 *Intentional ignorance*

The governing coterie or the whole IIC population may deliberately ignore any knowledge or ideas coming from the outside. This can be presented as a conscious choice of truth or some other value over the less valuable competition. However, this can also be an expression of fear that alien ideas will undermine the adopted principles of the IIC. The rejection of alien ideas occurs when there are competing ideologies, such as various religions or various philosophies of probability. Although ignoring rational arguments is anathema for scientifically-minded people, this approach is actually practical, rational and understandable. This is so because of the practical benefits that representatives of an ideology, for example, a political party or political ideology, can derive from victorious propaganda.

15.3.4 *Deliberate ambiguity*

One can disarm critics of IIC's claims by making the claims' empirical status unclear. This is used predominantly by religion and philosophy. The claims that God is omnibenevolent, motion does not exist, and there is no objective truth can be interpreted as empirically supported laws. Real-life examples have illustrated all of these. When critics propose counterexamples to these claims, philosophers and theologians escape into abstraction.

In the case of the subjective philosophy of probability, the deliberate ambiguity is based on the multitude of meanings of the word "subjective."

There is also a somewhat different type of ambiguity. Some theories present themselves as philosophical or scientific, depending on the circumstances. This applies directly to Marxism and both philosophical theories of probability—subjective and frequency. Religion historically (even quite recently, about one hundred years ago) made claims and commanded authority reserved for science nowadays.

15.3.5 *Transubstantiation*

According to [90],

> Transubstantiation [...] is, according to the teaching of the Catholic Church, "the change of the whole substance of bread into the substance of the Body of Christ and of the whole substance of wine into the substance of the Blood of Christ. This change is brought about in the eucharistic prayer through the efficacy of the word of Christ and by the action of the Holy Spirit. However, the outward characteristics of bread and wine, that is the 'eucharistic species', remain unaltered."

The essence of transubstantiation is that an object may completely change its character, although no empirical changes can be detected, and no such changes are asserted.

Two hoaxes, by Duchamp and Sokal (see [73, 87]), will illustrate the declarative powers within IICs other than religion.

Duchamp displayed a ready-made urinal at an exhibition and titled it *Fountain*. There is some evidence that Duchamp intended it to be a hoax (see [73]), but this is uncertain, rarely mentioned, and clearly ignored. The Wikipedia article reports that "In December 2004, Duchamp's Fountain was voted the most influential artwork of the 20th century by 500 selected British art world professionals." An IIC could turn an ordinary object into a masterpiece by a mere declaration. The process required the involvement of a prominent artist in an essential way—Duchamp was an excellent Cubist painter and made other artistic contributions. An unknown person could not have produced a hoax of a similar impact.

Sokal's hoax was a hoax only because Sokal declared it a hoax (see Sec. 15.1.5). If Sokal had not declared his article a hoax, the article would have been accepted as a genuine contribution to postmodern philosophy to this day.

The two hoaxes show the power of declaration. A simple declaration can fundamentally change an object's or publication's meaning and significance. However, the power resides with the ruling group at the top of an IIC, as the history of van Gogh's paintings shows (see [91]).

Probability is considered subjective in Bayesian statistics

because a group of prominent philosophers and statisticians declared it so. If a similarly powerful group of Bayesians declares probability to be objective, then probability will be miraculously transubstantiated into an entirely objective concept.

15.3.6 *Barrier to entry*

IICs guard their territory. Contributions to the ideology, especially the extreme ones, are accepted as genuine only if they come from recognized members of the community.

A "white-on-white" painting and a pile of bricks (see [92]) are art pieces and can be displayed in a museum only if recognized artists created them. An identical painting and a pile of bricks made by ordinary people have no value.

The claim that it is safe to jump from a skyscraper made in [31] is taken seriously enough to merit a refutation. A similar claim made by an ordinary person would be ignored or would earn the person the stigma of a mental disorder.

The value of the article published by Sokal (see [87]) depends on the standing of the author within the IIC (postmodern philosophy in this case). When Sokal revealed his hoax, he placed himself outside the IIC and removed his article from the list of significant contributions to the field.

Michał Heller—a scientist, a Catholic priest and a member of the Papal Academy—wrote in [93] that

> [...] there is a traditional theological doctrine, called "apophatic theology," reminding us that if we say something about God, we should be certain that we are only expressing our ignorance (docta ignorantia).

If an ordinary person expressed this thought, he would be identified as an atheist, and Catholics would ignore his views. Catholic theology contains some claims contradicting standard logic, such as that there is one God and the Holy Trinity and that the Book of Job agrees with God's omnibenevolence. On the other hand, the Catholic Church labels some similarly paradoxical claims heretical.

Martin Heidegger said (see [94, p. 141]) "[...] when philosophy makes itself intelligible, it commits suicide [...]". A similar claim made by a non-philosopher would be considered an expression of utter ignorance.

Once you become an insider, you can get away with outrageous contributions contradicting the IIC's ideology.

15.3.7 *Arcane knowledge*

An effective defensive mechanism is to develop knowledge or methods so sophisticated that it takes years for an average person to master them. Outside critics might not have the time, energy, patience, and possibly talent to study the arcane knowledge of the IIC. As a result, the masters of the IIC may brand potential outside critics as ignorant and refuse to engage with them in a rational dialogue.

Intellectual products of IICs become increasingly extreme due to pressure on contributors to distinguish themselves. The result is that fewer and fewer people can understand intellectual products, and the community is led by a few trendsetters who dictate the rules and award recognition.

Arcane knowledge exists in each of my examples of IIC's: science, mathematics, religion, art (or rather art criticism), and philosophy (especially in some of its subfields, such as German idealism). The creation of arcane knowledge is a natural and inevitable consequence of normal intellectual activity. An interesting or significant direction of thought will attract talented individuals who will develop the field far beyond the obvious and the ordinary sooner or later. Arcane knowledge becomes a problem when it results in intimidation via erudition. This is especially true in philosophy and art. The display of arcane knowledge frightens away potential critics and allows erudite authors to sell low-quality intellectual products as legitimate contributions to the field.

A social consequence of arcane knowledge is that it may take years to learn it to the degree sufficient to emulate it. Hence, the apprenticeship is the standard entryway into an IIC.

However, such an educational experience is rarely limited to learning purely technical skills and theoretical assertions. The apprentice is likely to learn the culture of the IIC and, therefore, join the IIC as a fully supportive member.

Arcane knowledge may require modifying or rejecting some of our basic instincts altogether. A well-known example is that of quantum mechanics. Some mathematical claims are equally counterintuitive, for example, the existence of non-measurable sets (and paradoxes based on these) or Gödel's theorems. Theological claims are often counterintuitive to outsiders. Counterintuitive claims are some of the barriers between an IIC and the outside world.

Arcane knowledge takes an unusual form in art. Some art pieces are indistinguishable from everyday objects, such as unfinished paintings, child paintings, etc. Examples include "white-on-white" and similar monochromatic paintings, ungrounded canvas cut with a knife, and an unframed mirror hanging among paintings at a major museum. It takes years to learn how to tell the difference between some of the paintings of Malevich, Mondrian or Kelly and experiments of an artistically uneducated and untalented individual.

Some IICs reduce their membership via exaggeration of ideological features so that fewer and fewer people (in terms of percentage) can consume and understand IIC's intellectual products. Fewer people can learn the rules and be able to tell genuine products from fakes.

15.3.8 *Stratification*

Some IICs have (at least) two intellectual levels: professional and popular.

Catholic theology, Marxism, and the subjective and frequency philosophical theories of probability have very sophisticated theories coexisting with their popular versions.

Professional ideology is developed by sophisticated intellectuals who do not want to be limited by anything in their intellectual endeavours, and so they generate arcane theories for their own pleasure and to impress their equally sophisticated colleagues. They create language and concept repositories inaccessible to ordinary people.

The two sides of an ideology are a marriage of convenience. The sophisticated side gets economic and practical support, including money, jobs, academic environment, etc. The popular side of an ideology needs the sophisticated counterpart to appear profound, thoughtful, well-justified in its ventures, mature, respectable and reasonable.

Discrepancies between the two sides of an ideology pointed out by outsiders are either totally ignored or discounted as a criticism coming from people who have their own agenda.

A type of ordinary hypocrisy exists in philosophy—theories jump between "naive epistemology" (NE) and "sophisticated epistemology" (SE) whenever convenient. Examples include the following.

Catholicism: (NE) Miracles always benefit humans. Hence, God is good. (SE) Multiple cases of genocide occurred without God's intervention. Nevertheless, God is good.

Communism: (NE) Workers in capitalist countries are oppressed and exploited. (SE) Workers in communist countries are the ruling class (despite obvious evidence to the contrary).

Relativism: (NE) Historical examples show truth distortions due to social or political pressures. (SE) Science does not come closer to the truth than other intellectual methods.

15.3.9 *Pliable logic*

Some IICs seem indifferent to or at least highly tolerant of logical contradictions. This is an effective defensive mechanism because it defuses a well-known and well-established method in ideological disputes. Illogical reasoning is typically reserved only for some aspects of the ideology, so the theoretical edifice appears solid.

Nobody wants to hold religion, art criticism and philosophy

to the same logical standards as those in mathematics. This license is sometimes tolerable—few Catholics or non-Catholics are bothered by the simultaneous claims that there is one God and the Holy Trinity. However, the license is often stretched beyond the reasonable. Non-Christians find it hard to comprehend how one can square the Inquisition and the Ten Commandments or how to reconcile God's infinite love and omnipotence with several cases of genocide.

The result of the selective logic is the loss of trust between the members of IIC and outsiders. This helps to maintain the siege mentality among the IIC members—the need to defend the ideology against disrespectful outsiders.

15.3.10 *Suspension of disbelief*

According to [95], "Suspension of disbelief [...] is the intentional avoidance of critical thinking or logic in examining something surreal, such as a work of speculative fiction, to believe it for the sake of enjoyment."

Religions typically include large numbers of beliefs that are entirely different from everyday experience. Philosophers tolerate claims in their field that challenge common sense, such as the claims of the non-existence of motion or objective truth. Mathematicians routinely use models of the space that have more than our three dimensions.

The suspension of disbelief goes deeper than just making claims contradicting our experience. In [31], Feyerabend effectively claims that jumping from a skyscraper is safe (see Sec. 4.1.3). This type of advice could be considered criminal in real life. Feyerabend can get away with it because of the implicit suspension of disbelief—everybody knows that Feyerabend cannot be serious.

If a doctor treats a patient and the treatment does not work, the doctor can be sued. A Catholic priest may be called to the sick person's bed and asked to pray for a quick recovery. If the prayer does not work, the priest is not liable.

The subjective philosophy of probability claims that there is no objective probability, so there is no rational reason to think that drinking increases the probability of a car accident. This provides another example of a claim that could be considered criminal if implemented as an offer to drink alcohol in a bar. However, subjectivists are in no danger of being prosecuted—the implicit suspension of disbelief is an effective shield. Nobody believes that subjectivists are serious with their claims, so nobody will ever take an author of a book promoting subjective philosophy to court for complicity in a DUI (driving under the influence) accident.

My examples show a particular form of the suspension of disbelief. People, including outsiders, recognize the specialized context of an IIC and adjust their expectations relative to normal life.

IICs erect barriers around their ideologies by introducing different standards of suspension in disbelief than those in the outside world. This makes some of their claims incommensurable with those of the outsiders, providing an effective defense.

15.3.11 *Sabotage*

Interestingly, ideologies as diverse as Catholicism and communism developed similar defensive mechanisms. Both ideologies employ sabotage as an explanation for unwelcome events. In the Catholic view of the world, all good events are due to God's beneficial nature. All misfortunes can be attributed to the devil. Communists celebrated all economic, scientific and social successes as results of the communist system of government. Many economic setbacks were attributed to saboteurs—capitalists, counterrevolutionaries, etc.

15.3.12 *Narrow focus*

Members of IIC learn to use their ideology only when convenient. Relativists claim that there is no objective truth, but they apply this philosophical claim only when it suits them; for example, the book [31] praised the Chinese Communist Party for suppressing Western medicine. Relativism is never applied to support racist behavior, for example, the racist behavior of some white juries. Subjectivists use their philosophy to support subjective priors in Bayesian statistics, but they do not support coherent opinions widely considered unethical, such as antisemitism.

15.3.13 *Oversimplification*

Our present culture suffers from the short attention span of individuals and mass media. However, attempts to condense and simplify the message started long ago. Several ideologies sell themselves using short slogans. The American system is represented as "freedom." Christianity is based on the commandment to "Love thy neighbor." Communism exhorts "Proletarians of all countries, unite!" The subjective philosophy of probability is based on the claim that "Probability does not exist." The frequency philosophy of probability can be summarized by "Probability is frequency."

While each slogan contains a grain of truth, the complete picture is much more nuanced than the slogan suggests. Critics of an IIC would go much further and say that the slogan points to the ideology's only bright spot and tries to hide multiple skeletons in the closet.

15.3.14 *Resurrection of functionality*

Marxists clearly and unequivocally declared that God does not exist. Yet, they created an ideology that was a religion for all practical purposes. Real communism in the Soviet Union under Stalin had

(a) God—Stalin; the target of prayers,
(b) Prophets—Marx, Engels, Lenin,
(c) Holy book—"Das Kapital" by Marx,
(d) Church (organization)—the communist party, as hierarchical and undemocratic as the Catholic Church,

(e) churches (buildings)—local communist party headquarters,
(f) rituals—parades, meetings, etc.,
(g) prayers—slogans displayed in public places,
(h) youth indoctrination—Komsomol youth organization and its practices,
(i) religious art—songs, visual arts, literature.

Subjectivists pronounced the death of objective probability. Yet, their implementation of subjective probability is almost identical to that used by the strictest objectivists. The differences are stressed for ideological reasons. The objective and subjective applications of probability have 99.99% in common. The objective probability was resurrected under the name of subjective probability or Bayesianism.

15.3.15 *Distortion by experts*

Experts invest so much time learning and developing technical tools for a given branch of intellectual activity that sometimes the techniques alienate themselves and become the main object of study or activity. Examples include mathematical techniques and artistic experimentation. At some point, nobody seems to care anymore whether the techniques are helpful and meaningful. IICs tolerate this highly specialized behavior, mainly inaccessible to outsiders.

15.3.16 *Deference*

There are numerous examples of deference to authority. Philosophers used to and continue, to some extent, to refer to the ideas of Plato and Aristotle. Well-known writers and politicians are often quoted. However, sometimes deference goes beyond the boundaries of the rational. Some well-known artists created art pieces displayed in museums that would have no value if ordinary people had created them. The same applies to philosophy. Some ideas, like "motion does not exist," are immortal. Many non-experts assume that if a prominent specialist expresses an idea, it must be profound, even if it sounds silly. I guess that many people are intimidated by the claim that "probability does not exist." They think it must be very profound since it sounds totally irrational.

15.3.17 *Loyalty and inertia*

Many maintain membership in an IIC out of loyalty or inertia. Loyalty can be considered a moral principle, while inertia is more practical. Changing the status of one's relationship with an IIC may require costly practical investment or considerable mental effort.

15.3.18 *Extremism*

Extremists push an IIC away from competition and criticism. While extreme ideological positions may be easy for outsiders to identify and criticize, the adoption and domination of extreme interpretations and actions provide immunity from outside influence. Extremists are often the loudest and most visible representatives of IIC. Their assertive behavior pushes rank-and-file members toward the extreme, and their self-confidence erects a hard barrier against outside critiques.

PART 4

Book reviews

Chapter 16

Latter Day Subjectivism

This chapter is devoted to the critical review of two recent books on the philosophy of probability by subjectivist authors. I have chosen [6] by Diaconis and Skyrms and [7] by Lindley for several reasons. First, the books are recent, so we can hold the authors to higher standards than those of earlier books—the authors had an opportunity to digest various arguments produced in support and against the subjective philosophy. Second, the books are aimed at a general audience, so they do not contain complex arguments. Third, the books are well-written. I enjoyed reading them just like I would enjoy reading a novel. The fourth reason is that the books contain many claims that I disagree with, providing me with many opportunities to present my own ideas.

There is also a personal reason. I know Diaconis personally. I attended several of his talks and had the chance to exchange mathematical and philosophical ideas with him on many occasions. Diaconis is one of the top probabilists and statisticians. He is also known for his parallel career as a magician and for excellent talks.

I knew Lindley. He might have been the most prominent among the fanatical supporters of subjectivism and the most fanatical among the prominent supporters of subjectivism. He was a visiting professor at Berkeley when I was a graduate student. The Statistics Department, where I was a student, was known for its frequentist bias. Lindley, a Bayesian statistician, visited the Operation Research Department instead. I took my first course on Bayesian statistics from Lindley. He was an excellent teacher—no sarcasm here.

The two books are very different in their structure and their intellectual goals. Diaconis' and Skyrms' book [6] contains a systematic review of every significant philosophical idea related to probability. Lindley's book [7] has the ambition to educate the general public on the subjective approach to probability but has no ambition of being a comprehensive introduction to the philosophy of probability.

Another significant difference between the two books is how they deal with inconvenient examples. Lindley explicitly says there is nothing wrong with unusual probabilistic opinions. His line of defense is the ostrich strategy. His examples of unusual probabilistic opinions are harmless. No practically significant unusual probabilities are discussed. Diaconis and Skyrms do not hide their discomfort with unusual probabilistic views, such as "dogmatic" prior distributions. They are not afraid to point out the irrational consequences of such opinions. However, they offer no solution to the conundrum, rejecting the only two promising ideas: Cournot's principle and the frequency interpretation of probability.

I must emphasize that my "review" of the two books is not a traditional book review. I discuss only selected passages and aspects of both books. My narrow goal is to show that modern subjectivists hold strong objectivist views.

I hasten to add that the logical contradictions and confusion that I found in [6, 7] are not any more profound than those in the writings of de Finetti (or von Mises, for that matter). Nevertheless, there are several good reasons, mentioned earlier in this section, for directing my criticism at the two books—they are recent, accessible, and well-written, so they are more likely to exert a strong influence on students of statistics and the general audience than the publications of de Finetti.

The reader may wonder why I have chosen two subjectivist and no frequentist books for a review. The reason is very simple—I am unaware of any modern philosophical book where the basic tenets of von Mises' philosophy are presented sympathetically to a broad audience. I do not know a book supporting the concept of a collective and claiming that individual events do not have probabilities. This contrasts sharply with the two books under review, where the basic ideas of the subjective philosophy of probability are presented and defended.

Before presenting long and exhaustive analyses of [6, 7], I will briefly discuss [96] and the introduction to subjective probability in [38] to show that the inadequate and misleading presentation of the subjective philosophy of probability is not limited to two recent publications aimed at the general public—it seems to be a perennial problem in professional philosophy as well.

16.1 Philosophy for professional philosophers

The book [96] by Skyrms presents the subjective philosophy of probability and its analysis. The language of the book and the choice of topics make it clear that the book is addressed to professional philosophers. Skyrms examines various concepts and problems related to probability in great detail. Chap. 2 is devoted to the Dutch book argument. As usual, it is argued that a rational decision-maker should use coherent probabilities to avoid the Dutch book. Chap. 3 is titled "Learning from experience." At the beginning of the chapter, Skyrms mentions the classical problem of induction using the term "uniformity," thus indicating that we are entering one of the most challenging areas in the philosophy of science. The bulk of the chapter is devoted to exchangeability. What I find missing is an explanation of what "learning" means. One may guess that one "learned" if the new state of one's knowledge is better than the old state of one's knowledge. The criterion of quality is not explained. I would guess that the "better" state of knowledge is closer to objective probabilities than the old state of knowledge, but subjectivists cannot adopt this definition of learning. Also, the book does not clarify why one should apply exchangeability. Assuming that exchangeability is objective and can be reliably recognized, what would a decision-maker lose by refusing to apply exchangeable models for an arbitrary reason?

The book [38] by Kyburg and Smokler is a collection of essays by prominent philosophers, including two essays by de Finetti. The introduction (see pp. 13–14) presents the following standard argument justifying coherence. A typical rational person would avoid a Dutch book, i.e., a collection of decisions that would result in a loss under some circumstances but never generate a net gain. Applying de Finetti's strategy, known as coherence and based on probability, eliminates all potential Dutch books. On p. 15, Kyburg and Smokler write (emphasis in original):

> In some ways, the most important concept of the subjectivistic theory is that of *exchangeable events*. [...] a way was discovered to connect the notion of subjective probability with the classical procedures of statistical inference.
>
> In general, the connection takes this form: it is shown that regardless of the subjective probability assignments with which a person starts (or rather, within very wide limits of such assignments), the results of applying the procedures of Bayesian statistical inference are the same: the same action is to be taken, the same hypothesis accepted, the same (or nearly the same) value of a parameter adopted.

This quote illustrates two important points. First, the authors sweep under the rug a fundamental assumption required of claims of this sort. For the results to be "the same," two statisticians must agree on the exchangeability (or lack thereof) of the data; otherwise, the claim is trivially false. Hence, one has to assume that data are objectively exchangeable (or not), independent of human beliefs, and statisticians can effectively recognize exchangeability. While the objective character of exchangeability is obviously (implicitly) assumed by statisticians, it is a major philosophical problem closely related to the famously hard problem of induction.

Second, Kyburg and Smokler do not provide any justification for using exchangeability. Suppose for the sake of argument that a statistician does not like exchangeability and never applies this notion to any sequence. Would their statistical conclusions be "far from the truth"? If yes, in what sense? Would the decisions be less beneficial than those when exchangeability is used? If yes, in what sense? The silence is especially striking because the justification for coherence is clear and empirically testable.

I find it rather disappointing that the introduction in [38] is quite misleading because the careful analysis of the subjective philosophy of probability presented in [42] shows that Kyburg had no illusions about this theory. In contrast to [38], [42, p. 176] says (emphasis in original),

> Given that two people have degrees of belief that satisfy the probability calculus, and given that their degrees of belief satisfy some relatively mild constraints in addition to the coherence constraints, then, as empirical evidence accumulates, their distributions of beliefs will become more and more nearly the same. Of course there is nothing *in the theory* that requires that even the mild constraints will be satisfied for a rational person.

Elsewhere, we find some very powerful words in [42]:

> [p. 158] [...] I shall argue that although the theory appears to be all things to all people, in fact it is a snare and a delusion and is either vacuous and without systematic usefulness, or is simply false.
>
> [p. 171] Subjectivistically interpreted, Normative Bayesian Decision Theory, whatever its heuristic virtues, is either philosophically vacuous or impotent.

My main point of disagreement with [42] is about "relatively mild constraints" in the first quote. These constraints are typically the assumption of observations' exchangeability. Exchangeability is one of the fundamental concepts in Bayesian statistics. At the same time, exchangeability presents a profound philosophical problem, as difficult as the problem of induction.

16.2 "Ten Great Ideas About Chance" by Diaconis and Skyrms

16.2.1 *General overview*

Ten Great Ideas about Chance has all the attributes of a classic. I am not sarcastic or hypocritical. The book is exceptionally well-written and accessible. Diaconis and Skyrms left no stone unturned. I am amazed by how much material Diaconis and Skyrms packed in a relatively short book compared to a typical philosophical monograph on probability. I am not new to the area, but I truly enjoyed reading the book and learned many facts I had not known before. I am impressed by the ability of the authors to present the main ideas of various theories in brief but accurate ways. Many readers will want to understand some of these ideas at a deeper level—Diaconis and Skyrms have reading suggestions for them. Many illustrations help explain technical points; other pictures add a human touch to formal arguments. Overall, the book is well written, accessible and, therefore, seductive to the point that it can become a standard philosophy of probability reference for a generation or two. I will show that, unfortunately, the book's philosophy is highly confused, so there is potential for significant intellectual damage.

16.2.2 *How to eat your cake and have it too*

The problem with [6] is the same as with all subjectivist books and subjectivism in general. Subjectivism is an attempt to create something out of nothing. The weaknesses of the subjectivist philosophy are hidden in different books in different ways. Diaconis' and Skyrms' book is a complex mixture of science and philosophy, so an unprepared reader may find it very hard to find a skeleton in the closet. In this section, I will outline the general picture. This will be followed by many sections with analyses of specific philosophical claims made in [6].

Science differs from mathematics and philosophy in handling its assumptions. Mathematicians and philosophers try to specify all possible assumptions that underlie their fields. In mathematics, the tradition goes back to Euclid. The axiomatic foundations of mathematics have been reexamined in the nineteenth and twentieth centuries. Philosophers have been aware of the importance of assumptions since antiquity. Pyrrho is credited with the creation of skepticism. A form of skepticism is based on the observation that every claim requires some assumptions, and all assumptions can be questioned. Any attempt to prove all assumptions will lead to an infinite regress.

Scientists explicitly state only assumptions relevant to the object or phenomenon at hand. For example, it is a law of nature that water boils at 100 degrees Celsius under normal atmospheric pressure. The assumption of the normal pressure is stated explicitly only when it is not implicit in the context. Science is based on the principle of induction, but no scientific paper or book ever states this assumption in an explicit form.

So far, I have pointed out a difference between science, and philosophy and mathematics; the difference is how explicitly the assumptions are stated. There is also another significant difference. In science, most of the time, the fundamental ontological and epistemological assumptions are ignored because they are considered obvious. Science is based on a tacit consensus that the physical universe exists and that there is objective truth. None of these claims is considered obvious in philosophy—these claims are the subject of mainstream philosophical investigations. Science does search for fundamental theories, such as the theory of evolution, quantum physics and relativity, but ignores its own philosophical foundations (the quantum theory being the exception that confirms the rule). Mathematics is in between. Some mathematicians almost entirely ignore the part of mathematics that deals with foundational issues (logic, set theory), while others focus on mathematical foundations.

Diaconis' and Skyrms' book mixes philosophy, mathematics and science to a degree rarely found in a book of any genre. Hence, it is easy to jump from one mode of thinking about assumptions to another. Chap. 2 on "Judgment" is almost entirely devoted to a philosophical justification of "coherence," that is, an attempt to derive the rules of probability from assumptions that many people would be willing to accept as "self-evident." Chap. 9 on "Physical chance" touches upon philosophical foundations of quantum physics but completely ignores the philosophical foundations of probability. The Dutch book argument features prominently in Chap. 2 but does not appear in Chap. 9.

This methodological diversity results in that "self-evident" assumptions are discussed only in the philosophical parts of [6]. The authors implicitly encourage the reader to accept the Dutch book argument in Chap. 2. The implicit assumptions of the argument are indeed easy to accept. However, the argument generates an extremely weak theory that says that any assignment of probabilities should agree with the standard rules of the mathematical theory of probability (effectively, only one rule—the additivity of the probability as a set function). Chap. 9 and the discussion of hypothesis testing starting on p. 115 are examples of typical scientific discourse, where philosophical assumptions are taken for granted and hence not worth discussing.

The above describes the general framework in which Diaconis and Skyrms make an impression of building a powerful theory based on very weak assumptions. Let us have a look at a concrete example. On p. 202, we read "... your probabilities and your conception of repetition of the same experiment are up to you. You and I may differ." Suppose the outside temperature is measured in Chicago daily at noon for ten years. Suppose that someone proposes to consider the measurements to be exchangeable. The historical difference between average temperatures in January and July in Chicago is about 30 degrees on the centigrade scale. No scientist would ever accept the exchangeable model because it completely ignores summers and winters in Chicago, a well-known fact. How can we interpret the claim that "... your probabilities and your conception of repetition of the same experiment are up to you" in relation to this example? There are two choices.

(i) Diaconis and Skyrms advocate unlimited tolerance. The exchangeable model for Chicago temperatures is consistent, just like many non-exchangeable models. Hence, all scientists should show tolerance and consider the exchangeable model as valuable as any other model. This kind of attitude is limited to pure philosophy. The fundamental point is that the Dutch book argument does not justify rejecting the exchangeable model for Chicago temperatures. The weak assumptions of the subjective philosophy cannot generate a conclusion that is considered mainstream in science.

(ii) Diaconis and Skyrms accept and advocate rational restrictions on beliefs about symmetry. This aligns perfectly with science as we know it. However, whatever rational limitations might be, they require assumptions stronger than those going into the Dutch book argument. The traditional extra assumptions are identifying frequencies and probabilities, and Cournot's principle. These are rejected on p. 66. No alternative justification for the widespread preference for some (consistent) probabilistic models over other (consistent) probabilistic models is given in [6].

The authors strongly support using symmetry in probability applications. This is exemplified by the extensive discussion of exchangeability in Chap. 7 and the following claim on p. 124: "...if you were worried about whether chances exist, you can forget your worries. *De Finetti has replaced them with a symmetry condition on degrees of belief*" (emphasis in original). One cannot derive restrictions on the symmetry of beliefs from the Dutch book argument. One can take them as an assumption or justify them using arguments based on identifying frequency and probability or Cournot's principle.

I strongly support symmetry; see my analysis of (L4) in Sec. 9.1.2. However, a naked symmetry principle will not work. It needs (L6), my version of Cournot's principle. It also needs a broader philosophical context—this is provided by my theory of "resonance" (see Sec. 7.2). Symmetry is the foundation of the logical philosophy of probability, also known as the "a priori" theory of probability, and it is plagued with interpretation problems in that context (see Sec. 2.2). The discussion of "grue" emeralds on p. 202 of [6] is a good illustration of problems with symmetry. An object is "grue" if it is green until 2050 and blue afterward. The philosophical problem is that the empirical evidence seems to support equally the claims that "all emeralds are green" and "all emeralds are grue." Most people do not find this conclusion palatable. These examples show that symmetry deserves more than emotional support on the philosophical side.

Diaconis' and Skyrms' book provides a meticulous and detailed discussion of different levels of skepticism in Chap. 10. The authors point out that one cannot prove that any of these levels is correct. One has to choose. This is fine with me, except that the authors jump from one level to another according to what is most convenient in a given chapter of their book. Their positive philosophical attitude towards the Dutch book argument and the negative philosophical attitude towards identifying frequency and probability, as well as Cournot's principle, cannot provide the basis for the scientific models they support.

16.2.3 *Verifying probability values*

My main criticism of [6] is that it embraces subjectivism only when convenient. This section will show how [6] adopts the subjectivist position. The following section will be devoted to the book's objectivist elements.

Justifying a procedure for assigning probability values is a philosophical problem harder than the famous problem of induction because the latter is a special case of the former, when all observations, potentially taking different values, are the same. Very few solutions to the philosophical problem of probability were proposed. Each one was criticized and rejected by some philosophers and users of probability. This creates a temptation to cut corners by declaring that the problem does not exist. This is the root of de Finetti's claim that "Probability does not exist."

Science as we know it is based on verification. If a scientific claim is made or a theory is propounded, verifying it before it can be generally accepted is necessary. In Popper's version of the same idea, one must try to falsify the claim or theory, but this is not an essential difference. The point is that scientific claims have to be compared with empirical observations. The reason for such a requirement is self-evident—without strict verification standards, anyone and everyone could make arbitrary claims. In practice, many people do, especially outside science. But even on the fringes of science, such as "alternative medicine," dubious businesses prey on the uneducated or the desperate and make money selling snake oil.

Only three clear ideas were ever proposed by philosophers (or scientists interested in philosophy) for verifying probability values. They are

(i) Equating frequencies and probabilities.
(ii) Cournot's principle.
(iii) The Dutch book argument.

Diaconis and Skyrms discuss all three methods in their book. On p. 66, we read, "The argument that the law of large numbers solves the inverse problem is a fallacy." The "inverse problem" means going from data to the probability value. On p. 65, [6] says that "Bernoulli argued that [...] it would be *morally certain* that relative frequency would be (approximately) equal to chance" (emphasis and parentheses in original). The authors believe that the observed relative frequency cannot be considered close to the probability of an event. It seems that logic inevitably leads to the conclusion that if someone expressed an opinion that the chance of an event is p and later the observed relative frequency happened to be considerably different from p then, according to Diaconis and Skyrms, this could not be considered a falsification of the belief that the chance was p.

Cournot's principle is discussed on p. 66 of [6]. The authors point out that Cournot's principle is essentially the same as Bernoulli's "moral certainty argument" presented on p. 65. The authors say on p. 66 that "Cournot's principle, taken literally, is absurd." I have not found any "non-literal" version of Cournot's principle in the book presented approvingly. Hence, we can conclude that Diaconis and Skyrms unconditionally reject Cournot's principle.

The Dutch book argument is introduced on p. 25 and discussed in detail throughout the chapter. The authors are supportive of the argument. One cannot stress enough that the occurrence of a Dutch book only shows that the family of probability values used for the relevant decisions were inconsistent;

that is, they did not satisfy one of the mathematical properties of probability (Kolmogorov's axioms). This method is never used for verification purposes in real life because direct verification of the mathematical correctness of a probabilistic model is much more practical than the search for a potential Dutch book.

I conclude that Diaconis and Skyrms reject (i)-(ii) and accept (iii). In other words, they appear to be strict subjectivists or, in their own words, "thorough Bayesians" (see their Preface).

16.2.3.1 *Utility of probability*

The problem of utility of probability is closely related (although not identical) to the problem of verifying probability values.

I think that the degree of Seattle sunsets' beauty is 25. Is this number useful in any way? What could we gain by assigning numerical values to our opinions about beauty?

If someone says the probability of rain tomorrow is 0.25, is this number useful? There seems to be a broad agreement that the answer is "yes," yet there are very few justifications for the affirmative answer provided by philosophers. They are

(i) Objective probability exists. Knowing the truth has value irrespective of any practical benefits.
(ii) Knowing objective probability can result in beneficial decisions and, therefore, beneficial outcomes. In this context, frequency interpretation is often used. Sometimes, Cournot's principle is invoked.
(iii) The Dutch book argument is typically applied to support subjective probability but equally applies to objective probability.

There are good reasons to believe that the authors do not believe in objective chance. They write on p. 200, "De Finetti, like Hume, believes that there is no such thing as chance in the world and shows that we can have the virtues of Bayes' analysis without the baggage." There are many other hints throughout the book that objective probability does not exist or is not needed.

Assuming I am right, options (i) and (ii) are eliminated. So (iii) remains the only reason why probability is beneficial. Once again, we are tempted to conclude that the authors are strict subjectivists or "thorough Bayesians."

16.2.3.2 *Quantum vacuum*

I know quantum physics at the level of popular science books. Even by quantum physics standards, the role of vacuum is strange. In quantum theory, the "empty space" plays a significant role and does not appear empty at all. I use this simile to emphasize the role of silence in philosophical writings. Silence can carry a powerful message, for example, when politicians do not comment on current events. Silence in philosophy can be an equally powerful way of expressing ideas, but often, silence is hard to notice because it is hidden in the thicket of arguments.

An excellent example of meaningful silence in philosophical writings is the total lack of real-life examples of a Dutch book. The reader is offered artificial examples and is likely to subconsciously assume that "real" examples exist but are not presented only because they are too complicated or dull.

I have argued that Diaconis and Skyrms rejected in [6] the identification of frequency and probability, and Cournot's principle. I will show below that the identification of frequency and probability is implicit in many authors' arguments. The reader may conclude that the philosophical position of [6] on these issues is unclear. I claim that silence is a crucial piece of evidence.

The identification of frequency and probability, and Cournot's principle may appear to be easy and intuitive ideas to some people, but they are challenging philosophical problems. Identifying frequency and probability generalizes the induction principle, which is well-known as a tough philosophical problem. Recall that von Mises' philosophy supports the identification of frequency and probability at the cost of introducing collectives, a very awkward concept to deal with in theoretical mathematics and statistics, and applied statistics. My justification of Cournot's principle, embodied in (L6), takes considerable space in Sec. 9.3.

If the book [6] was inclined to identify frequency and probability or accept Cournot's principle, we would expect to see substantial arguments supporting these ideas. There are none. Compare this to the extensive discussion of the Dutch book in Chap. 2 of [6]. If the reader doubts the "official" position of [6] on the identification of frequency and probability, and Cournot's principle, the silence provides unambiguous proof that the authors reject both philosophical ideas.

16.2.4 *Objectivism leaking into subjectivism*

A major problem with [6] is its love-and-hate relationship with objective probability. I will show that the book has numerous claims, suggestions and philosophical positions that betray objectivist predilections. I think the logical inconsistency of the subjectivist-objectivist position is strikingly clear.

16.2.4.1 *Measuring the nonexistent*

Professor Kemikos has recently published a book presenting a new method for measuring the mass density of phlogiston. At the same time, Professor Fisikos published a book that developed a new method for measuring the speed of luminiferous aether relative to the Earth. It would never occur to me to criticize the two new measurement methods. According to well-established scientific theories, phlogiston and luminiferous aether do not exist (see [51, 52]). I would criticize professors Kemikos and Fisikos for wasting time on false phlogiston and luminiferous aether theories. What is the point of criticizing the two professors for developing experimental methods for measuring non-existent substances?

In a different area, I would be surprised to see an atheist philosopher criticize a Catholic theologian for arguing that Catholics should not believe in the Holy Trinity. If the atheist philosopher does not believe in God at all, what is the point of discussing arcane points of theology?

A section of [6] starting on p. 64 is devoted to "Bernoulli's swindle." On p. 65, we read, "...if frequency equals chance, then chance equals frequency. So, the argument goes, we have solved the problem of the inference from frequency to chance. This is Bernoulli's swindle." I suppose I am not isolated in thinking that "swindle" is quite a strong word. Why are Diaconis and Skyrms so emotional? They declared themselves to be "thorough Bayesians" so we are led to believe that they do not believe in the existence of objective chance. Therefore, their emotional attitude to Bernoulli's method of measuring chance is astonishing. If objective probability does not exist, then no method of measuring it can be better or worse than any other method.

On p. 67, the authors explain that the Bayes theorem should be used to find the probability given the frequency. This is fine, except that the only role of the Bayes theorem is to make the probabilistic beliefs consistent, and the only reason given by subjectivists for having consistent opinions is that they eliminate Dutch books. Bernoulli's "swindle" is so widely used in science that it has the status of religious dogma for much of the scientific establishment. If the swindle is so widespread, there must be numerous cases of Dutch books that scientists unaware of Bernoulli's fallacy suffer. No real-life cases of a Dutch book were ever reported in the book under review or any other publication I know of.

If a Bayesian statistician uses the "uniform prior" to analyze a sequence of one hundred tosses of a deformed coin, the posterior probability of heads will be almost equal to the observed frequency of heads. There is only one way to change this posterior opinion—change the prior distribution. The authors discuss only two types of priors different from the "uniform." One of the alternative priors affects only the probability of an extremally biased coin (it always falls heads up or always falls tails up). The other type are called "pigheaded" (see the next section for a discussion of these priors). Clearly, Diaconis and Skyrms used another strong word, "pigheaded," to discourage readers from using these priors. The authors do not explain what other alternative priors should be used and under what circumstances. The lack of examples leaves the strong impression that the uniform prior should be universally used. But then probability is frequency.

The only way the claim that probability is not equal to frequency can be made believable is to discuss a handful of real applications of Bayesian statistics in science where the prior distribution was not uniform and the median of the posterior distribution (i.e., the estimate of the probability) was different from the observed frequency. Without real examples, the claim that probability is not frequency sounds hollow.

16.2.4.2 *Pigheaded priors*

The section on "Consistency" (p. 114) discusses a "pigheaded" prior distribution and suggests that one should not use it. First, I have to warn the reader that the term "consistency" in the title of that section has a different meaning than that adopted in this book (and much of the philosophical literature on probability). In that section of [6], consistency is a property of the prior distribution that makes the posterior distributions converge to the actual physical chance. If you repeatedly toss a coin that is biased in a way unknown to you, your initial opinion about the probability of heads may be expressed as a distribution on the interval $[0, 1]$. An example of a "pigheaded" prior distribution is a distribution that gives no probability mass to values less than $1/2$. In other words, it is pigheaded to assume with certainty that the probability of heads cannot be biased towards small values. If the prior distribution is not pigheaded in this or similar sense (that is, it is "consistent"), then standard Bayesian arguments ensure that, when you observe consecutive tosses of the coin, your opinion will converge to the true physical chance of heads. Fine, except that this whole section of [6] is addressed to people who believe in objective physical chance. There is no discussion of what subjectivists may gain from avoiding pigheaded prior distributions. A pigheaded prior distribution protects its holder against Dutch books just like any other prior distribution. On top of that, the convergence of relative frequencies seems to be taken as an empirical fact. According to some consistent probabilistic views of coin tossing, the relative frequencies will not converge at all. The unstated belief in the convergence of frequencies betrays an objectivist and frequentist view of the world.

On p. 199, the epithet "dogmatic" is used for the prior, formerly called pigheaded in the book. The argument against this prior is that the frequency of heads in a long sequence of tosses of a coin may converge to 1/4, but the opinion about the future probability will always stay above 1/2. The terms "pigheaded" and "dogmatic" are pejorative, so it is clear that Diaconis and Skyrms do not like these types of priors. However, the only argument against them is that they may result in divergence of the posterior probabilities and observed frequencies. It seems that [6] takes an *extreme* frequentist position here. The book considers the convergence of posterior probabilities and observed frequencies a self-evident benefit because the Dutch book argument does not justify it, and no other justification is given.

The classification of priors into skeptical and dogmatic on p. 199 does not seem to agree with the philosophical preferences expressed in the section on "Finite exchangeable sequences" starting on p. 125. The authors show how one can use mathematics to eliminate infinite sequences from philosophical arguments. It is evident that all sequences ever observed were and will be finite. It is a philosophical choice whether we should consider idealized infinite sequences. If we limit our considerations to finite sequences, skeptical priors are no better than dogmatic priors. A skeptical prior may have an extremely small density on most of the interval $[0, 1]$. It will act as a dogmatic prior for all practical purposes.

The discussion of dogmatic priors is limited to a rather abstract setting. Coin tossing is more a mathematical model than a real-life example. Many people hold unshakable beliefs—religion, ethnic prejudices and conspiracy theories are based on strong beliefs ("priors") of their followers. I am missing in [6] a discussion of the rationality of such strong beliefs. The coin tossing example is highly misleading. People with no experience with statistics and other sciences may not realize that dogmatic priors encapsulate real-life ideologies well despite being an abstract example. Paradoxically, statisticians and scientists may have the same attitude because pigheaded priors never occur in scientific practice. Both groups might miss the real-life examples—religion, ethnic prejudices and conspiracy theories. The book does not make it clear whether Diaconis and Skyrms accept highly concentrated real-life priors as totally rational (because they are coherent) or whether they believe in some principle that makes these priors irrational.

16.2.4.3 *Frequentist attack on frequentism*

"Why most published research findings are false" is the title of the section starting on p. 115. The section is devoted to the poor performance of frequency-style hypothesis testing. In the last paragraph of the section, the authors assert that "The proper guidance is Bayesian" to find a way out of the conundrum. The authors do not seem to realize how profoundly frequentist their arguments are. They start the section by reporting a greatly disappointing fact that less than half of the results published in leading psychology journals could be replicated. This observation and the rest of the section clearly show that the authors believe that the replication rate is lower than it can and should be, and some methods could improve (raise) the replication rate. The core of the argument is deeply objectivist and frequentist. The authors believe in the ability of scientists to develop and implement probabilistic methods that will yield observable and verifiable results—the improvement of the long-run replication rate. There is no discussion of the Dutch book and no indication that Dutch book situations arise due to a low replication rate. Several practical suggestions at the end of the section are clearly objectivist. A subjectivist can only propose methods to eliminate Dutch book situations. According to subjective philosophy, the only practical way to undermine a probabilistic method is by showing that it generates Dutch books. Subjectivists cannot claim that their practical suggestions will change long-run frequencies of replicable results because whatever consistent distributions they implement in their arguments, other consistent probability distributions predict different long-run frequencies. According to de Finetti's theory, no consistent distribution is closer to the truth than any other distribution because there is no "true" distribution.

Let us reexamine the authors' disappointment that less than half of the results published in leading psychology journals could be replicated. Why should anyone expect anything better than that? The philosophical foundation of replication is the principle of induction. So, the authors implicitly show their recognition and reliance on the induction principle, which is reputed to be hopelessly intractable among epistemologists. But wait, can we really use the induction principle in this context? We are discussing research results based on hypothesis testing, so the deterministic version of the induction principle cannot be applied. Its probabilistic version is called frequentism.

16.2.4.4 *Probability over chances*

I will present and discuss some quotes from pp. 118–119.

"[Bayes'] key insight was that one has to put probabilities on the chances and update them by conditioning. This gives us judgment about chances, informed by the evidence of frequencies." The statement strongly suggests that Diaconis and Skyrms believe in objective physical chance. Otherwise, why should anyone care about "judgment about chances, informed by the evidence of frequencies"?

"Different prior beliefs can approach each other when updated with the same evidence of a substantial number of trials." As stated, this is undoubtedly true, but the key word is "can." Two parties may have conflicting interests; for example, a pharmaceutical company may want to generate big profits while a consumer group may want to protect the population from harmful drugs. If there are absolutely no rational restrictions on the prior distribution, the two parties may choose prior distributions so different that no amount of data that can be conceivably collected over one thousand years will make "different prior beliefs approach each other." So either the quoted statement is irrelevant in many (most?) cases when it matters, or one has to assume that there are rational restrictions on prior distributions. I consider such restrictions equivalent to the existence of objective probability, but if the reader is allergic to the concept of "objective probability," then I will settle on "rational restrictions."

Here is another passage. "In some cases, with a large number of trials, Bernoulli's swindle gives approximately correct results [...]." The appearance of the word "correct" is surprising. A measurement of an objective physical quantity, such as the Moon's mass, can be "approximately correct" or not. With no objective probability, how can any method give an "approximately correct" result?

16.2.4.5 *Unwelcome surprise*

On pp. 120–121 of [6], the authors discuss tosses of a tack. They analyze the consequences of adopting the uniform prior for the probability of "point up" on a single toss. They show that if there are no points up in the first ten trials, there will be a point-up result in the subsequent ten trials with a probability very close to 1/2. Moreover, they show that for large n, the same is true: if there are no points up in the first n trials, there will be a point up result in the subsequent n trials with a probability very close to 1/2. They conclude the paragraph with, "If this is an unwelcome surprise, then perhaps the uniform prior is suspect." Why unwelcome? Can a Dutch book arise? Obviously not, because it never does when the model is constructed according to the standard rules of probability. So why unwelcome?

In the next paragraph, the authors discuss a prior that might be less "suspect." It gives probability mass 1/3 to 0 and mass 1/3 to 1. It spreads the rest of the probability mass uniformly over the interval $(0, 1)$. With the new prior, if there are no points up in the first n trials, then the probability that there will be a point up result in the following n trials will diminish considerably relative to the case of the uniform prior. What have we achieved by replacing the prior? The new prior did not eliminate any Dutch books because there were none to start with. I see only one possible explanation: The new prior gives better predictions in practice. However, to justify this claim philosophically, one needs one of the objectivist arguments, such as the identification of frequencies with probabilities or Cournot's principle.

16.2.4.6 *Markov fetters*

On p. 131, the authors start a section on "Markov chains." The authors ask, "What if your degrees of belief are not exchangeable?" The section briefly discusses Markov exchangeability, an example of a (consistent, of course) probability model. Markov exchangeable processes are mixtures of Markov chains, just like exchangeable sequences are mixtures of i.i.d. sequences (under reasonable assumptions). The theory of Markov chains (generally, Markov processes) is much more complicated and challenging than the theory of i.i.d. sequences. What is the point of using complicated models if one can use simple models? Recall that every consistent model will protect the decision-maker against a Dutch book, so both an exchangeable model and a Markov exchangeable model are equally capable of eliminating Dutch books.

If I believe blue is more elegant than orange and I buy a

house painted orange, I might be willing to spend some money on repainting it blue. Do scientists use Markov chain models instead of simpler i.i.d. models to satisfy their subjective feelings about probability? The amount of money spent on research in probability is enormous. Why is all this money wasted on scientists' personal feelings?

I find it hard to believe that the expression "your degrees of belief" refers to arbitrary beliefs. If this is the case, then Markov models should be abandoned—change "your degrees of belief" to save money. Otherwise, "your degrees of belief" must have a link to reality. The link is the objective probability, no matter how hard one may try to rename it or hide it behind philosophical sophistry.

16.2.4.7 *Is physics objective or subjective?*

Chap. 9, on "Physical chance," contains an excellent review of how probability enters physics: via thermodynamics, ergodicity, chaos and quantum mechanics. I do not see anything controversial in the presentation, except that the chapter leaves an impression that the authors are one hundred percent objectivists. There is no suggestion that it is OK to have arbitrary consistent probabilistic opinions that differ from the mainstream science. Probability in physics is presented as usual; well-understood theories are presented as facts; less understood theories are flagged as such, and the related difficulties are discussed.

Consider, for example, this sentence on p. 184: "Suppose the measurement axes A, B and C are chosen independently with equal probability by some local quantum randomizer, as indeed they are in one of the rigorous experimental tests of Bell." A reader who is not (over)sensitive to the issue of subjective versus objective probability, like I am, will certainly interpret "as indeed they are" as an objective factual statement about the equipment and not an arbitrary consistent opinion about probabilities held by some scientists.

16.2.5 *Useless symmetries*

The following thoughts can be found on pp. 201–202: "We emphasize the subjective nature of the symmetry. You and I may have different conceptions of the repetition of an experiment, since we may have different symmetries in our degrees of belief" and "... your probabilities and your conception of repetition of the same experiment are up to you. You and I may differ." This is a recipe for killing science as we know it. If scientists may arbitrarily choose "symmetric" sequences, they can perform an arbitrary sequence of experiments supporting an arbitrary scientific theory. My claim extends to the organization of society. Every society is based on the consensus on symmetries. Should we tolerate the (hypothetical) claim that women and criminals form a single exchangeable group?

The discussion of the frequency-style hypothesis testing starts on p. 115 with the dissatisfaction with the past replication rate and ends on p. 117 with some suggestions for improvement. If indeed "your conception of repetition of the same experiment [is] up to you", then frequentists may say that they consider the combined family of the past and future hypothesis tests to be non-exchangeable. Moreover, they may believe that the past and future hypothesis tests are independent. Hence, the low replication rate observed in the past does not say anything about the replication rate in the future. Frequentists may believe that, with a high probability, the future replication rate will be very high and, therefore, there is no need to improve publishing practices. I wonder whether Diaconis and Skyrms would consider such a hypothetical frequentist opinion rational and acceptable.

The remark "... your probabilities and your conception of repetition of the same experiment are up to you. You and I may differ." is highly misleading because it can be interpreted in a way that perfectly agrees with the current scientific and general practices, or it can agree with de Finetti's philosophy, but not both. Inexperienced readers will subconsciously choose the first interpretation and thus will swallow the bait. Temperature is considered an objective quantity, but people experience it in their own subjective ways. However, this does not mean that these personal sensations are totally arbitrary. Suppose two people decide to see a Spielberg movie at Superstar Cinema. In that case, one may expect a good experience because Spielberg is her favorite director, and the other person may look forward to seeing the movie at his favorite movie theater. They care about different symmetries, although both symmetries objectively exist (Spielberg's movie-making style and Superstar Cinema ambiance). In both examples, personal symmetries are strongly related to reality, although in different ways. This relationship with reality needs a philosophical explanation or justification, just like the principle of induction does. In de Finetti's theory, no symmetry can eliminate Dutch books more effectively than any other symmetry, so no justification for using (objective) symmetries is needed. In real life, the freedom to choose your own symmetry arbitrarily is an illusion.

16.2.6 *Sensational philosophy*

On p. 124, we read, "...if you were worried about whether chances exist, you can forget your worries. *De Finetti has replaced them with a symmetry condition on degrees of belief.* This is, we think you will agree, a philosophically sensational result" (emphasis in original). Sorry, I do not see anything "sensational" here. I consider the theorems of Bayes and de Finetti (the latter linking exchangeability and i.i.d. sequences) very elegant and useful in *mathematics.* The Bayes theorem and the concept of exchangeability are unquestionably useful in *science,* specifically, in Bayesian statistics and its applications. However, de Finetti's theory is sensational in *philosophy* only in the sense of being a sensational failure. If the "symmetry condition on degrees of belief" did not have to be related to the real world, that is, if people were permitted and encouraged to apply (or not) this condition to any sequence whatsoever, then the results would be disastrous.

Exchangeability is useful because humans can link their exchangeable opinions to appropriate real sequences. De Finetti's Dutch book argument does not explain why recognizing truly exchangeable real sequences might be beneficial. You can avoid a Dutch book even if you never apply exchangeability. Moreover, de Finetti put forward explicit arguments against objective symmetries.

The word "sensational" indicates how much Diaconis and Skyrms are focused on frequencies despite all their efforts to distance themselves from frequentism. Even if de Finetti's philosophical idea is "sensational," it is limited to sequences variously known as i.i.d., exchangeable or collectives. The review of partial exchangeability on pp. 131–133 of [6] only reinforces the impression that exchangeability is a concept with very limited applications. See Sec. 14.6 of the present book for a review of applications of probability that do not involve collectives. De Finetti's supposedly sensational idea has nothing to say about these.

16.2.6.1 *Exchangeability versus i.i.d.*

I will use this occasion to discuss a subtle philosophical point. Bayesians flatter themselves on inventing a concept that simplifies the foundations of probability. Apparently, all they need to do (simple) statistics is exchangeability: "*De Finetti has replaced them with a symmetry condition[...].*" The implicit claim is that there is just one symmetry condition instead of what is needed to apply the frequentist concept of i.i.d. (independent identically distributed) random variables to the same class of real sequences. A frequency statistician must recognize two features of the sequence—(i) independence of its elements and (ii) identity of the distributions of the elements.

It is true that to use exchangeability, one has to recognize only one symmetry, specifically, the invariance under permutations. However, this hides the fact that one has to be able to tell the difference between the following two cases: (i) the sequence is exchangeable, and (ii) the distributions of the sequence elements are equal, but the sequence is not exchangeable. To see my point, consider predictions for rainfall in Seattle during the first two weeks of May 2050. From the current perspective, the rainfall predictions (probabilities) for individual days are identical or very close to identical. However, the sequence of the 14 consecutive rainfall measurements is definitely not exchangeable. If it had been, the weather prediction business would have been entirely different from its current model.

Frequentists have to choose between the following models for the data:

(a) independent and identically distributed,
(b) independent but not identically distributed,
(c) identically distributed but not independent,
(d) neither independent nor identically distributed.

In practice, Bayesians can easily distinguish between (a)-(d) and find the right model. But the subjectivist theory pretends that Bayesians have to choose between two options only:

(1) exchangeable,
(2) non-exchangeable.

Consider a sequence of observations made in the same laboratory at different times. One of the observations is in the past lightcone of the other observation for any pair of observations. Suppose that someone asserts exchangeability of the sequence that is, the invariance of the joint distributions under the change of the order of the recorded values. This assertion contains an implicit claim of the lack of physical signaling, i.e., physical independence. Recognizing an i.i.d. sequence does not require resonance skills beyond those needed to recognize an exchangeable sequence.

16.2.7 *Sloppy terminology*

The first problem with [6] is its fifth word, more accurately, the fifth word of the title ("chance"). I rush to admit that *Ten Great Ideas about Chance* sounds great. However, the title sows confusion in the minds of unprepared readers. Inside the book, the authors follow the convention popular among some philosophers of probability, to use the word "probability" in reference to subjective probability and to use the word "chance" to denote objective or physical probability. The book's ten chapters are devoted to both interpretations of probability, not just objective chance. I realize that the more accurate title "Ten Great Ideas about Chance and Probability" could cut the readership by half, but the brevity and appeal of the title have their price.

The terminological confusion continues in the Preface, although I admit this might be my own confusion. The authors declare themselves to be "thorough Bayesians." I cannot tell the difference between "thorough Bayesians" and "ordinary Bayesians." The term "Bayesian" has multiple interpretations. In statistics, a topic raised in the book under review, it refers to followers of "Bayesian statistics," as opposed to statisticians who prefer "frequency" methods. The authors embrace a strongly anti-frequentist position in the section starting on p. 115 (they criticize frequency-style hypothesis testing). Many, perhaps most, statisticians have little, if any, interest in philosophy, so being strongly anti-frequentist in the field of statistics does not necessarily imply anything about a person's philosophical views. On the other hand, the term "Bayesian" is often considered a synonym of "subjectivist." Hence, "thorough Bayesians" might mean "strict subjectivists." I guess that the authors say that they believe, with de Finetti, that "probability does not exist."

16.2.8 *No one is betting*

On p. 26, we find, "Of course, there is no bookie, and no one is betting. Still coherence, like consistency, seems like a worthwhile standard." Here "coherence" refers to what I call de Finetti's consistency, and "consistency" means "logical consistency."

I guess that the amount of money spent on probability worldwide is in billions of dollars if we include salaries of all statisticians, scientists and engineers using probability, teachers of probability at high schools and colleges, office and college buildings where probability is calculated or taught; computers and software, etc. Is all this money spent just to maintain "a worthwhile standard"? I wonder how many business people believe their companies spend money on probability and statistics to support "a worthwhile standard." If the only benefit of probability is "a worthwhile standard," then probability should be abandoned and billions of dollars saved. Otherwise, it is an intellectual challenge for philosophy to explicate the benefits of probability. Eliminating Dutch books is a red herring.

16.2.9 *Are impossible events possible?*

I will address the validity and power of Cournot's principle because it plays a fundamental role in my system (L1)-(L6). Diaconis and Skyrms discuss Cournot's principle on pp. 66–67 and claim that "taken literally, [the principle] is absurd." The principle says that "small-probability events should be taken to be physically impossible." The authors present an example of a dart thrown at a target. Hitting a specific point has a minuscule probability. The principle suggests that hitting a point is physically impossible. Later, Diaconis and Skyrms consider a modified version of the claim, saying that hitting a point singled out in advance is physically impossible. I do not understand why the authors consider the claim to be "absurd." The standard model Bayesian and frequency statisticians use for this and similar experiments assumes that the hitting distribution has a density. For the game of darts, two natural distribution choices are uniform and two-dimensional normal. According to these mathematical models, the probability of hitting a fixed point is zero. In practice, it is impossible to determine whether a point was hit with infinite accuracy. We must agree on a procedure to check whether a "real" point is hit. Then, the probability of such an event is strictly positive but very small. Suppose that a person bets \$100 against one dollar that the point will not be hit. Is this irrational? Making this bet agrees perfectly with the standard decision theory adopted by Bayesians (see Chap. 2 of [6]) and with popular choices of subjective probabilistic opinions.

I think that the dart example was a poor choice to start with. It is hard to determine whether Diaconis and Skyrms talk about theoretical infinitely small points or physical points. The focus of the philosophical analysis is shifted, consciously or subconsciously, from Cournot's principle to the relationship between theoretically infinitely small objects and their real counterparts. Using a discrete example, such as the following, would be much better. Toss an ordinary coin 100 times. This experiment is easy for anybody to perform in under half an hour. Recording the results poses no practical or theoretical problems. The probability of 100 heads is 2^{-100}, which is smaller than 10^{-90}. There is no question that this event can occur according to all standard scientific and philosophical theories. Cournot says this event should be considered impossible. The largest private fortune is now smaller than 10^{12} dollars. Hence, betting all their fortune against one dollar that there will not be 100 heads is rational for every individual. That is all that Cournot says. Is this really "absurd"?

Let us consider a more practical example—the possibility of being hit by a meteor. Yes, I said "practical." There is no question that meteors fall to the surface of the Earth. Anyone hit by a big or medium-sized meteor will be instantly killed. One could protect oneself against small and medium-sized meteors by spending more time in buildings with concrete ceilings (for example, office buildings) and less time outside and in family homes. People consider the probability of being hit by a meteor so small that they totally ignore it. This illustrates how Cournot's principle is interpreted in real life.

The last two examples suggest the following interpretation of Cournot's principle: Some probabilities are so small that performing rigorous mathematical calculations in the decision-theoretic framework is a waste of time. In such cases, intuition can efficiently and reliably ascertain the calculations' results.

Cournot's principle is analogous to many other situations when we consciously or subconsciously ignore tiny physical quantities. We would not take any action if we learned that one poisonous molecule was in our food. We would ignore the temperature change if one air molecule in our room doubled its energy (from cosmic radiation). Why is it irrational to ignore extremely small probabilities?

Cournot's principle is practical advice. Examples show that it is easy to implement, and people routinely do so. Turning it into a philosophical (epistemological) proposition may be challenging, but denying its force means closing one's eyes to reality.

16.2.9.1 *Cournot's principle and Dutch book*

The rejection of Cournot's principle contradicts the Dutch book argument, wholeheartedly embraced in [6]. Recall that according to the Dutch book-based justification of probability, one should accept a bet B if and only if the bet's expected payoff is positive. The expected payoff or, better, the expected utility $E(U_B)$ of a bet is the sum of terms of the form

$$\text{(Probability of } A\text{) times (Utility of } A\text{)},$$

extending over all disjoint events A, such that the utility has the same value for all outcomes favorable to A.

If $P(A)$, that is, the probability of A, is very small, then the expected utility $E(U_B)$ will hardly change if we replace $P(A)$ with 0. Hence, pretending that A cannot happen will not change our decision. In other words, Cournot's principle will be implicitly applied.

There are two exceptional cases when replacing $P(A)$ with 0 can affect the decision.

The first case is when the change in utility is extremely large if A happens. Examples include loss of human life, war, nuclear plant accident, and natural disaster such as earthquake. However, in most cases like these, specifying the loss of utility is very hard, so even if $P(A)$ is known with great accuracy, it is very hard to estimate $E(U_B)$.

The second case when replacing $P(A)$ with 0 may affect the decision is when $E(U_B)$ is close to 0. But then the decision-maker is essentially indifferent towards placing the bet. Moreover, the uncertainty about the values of all probabilities and utilities of all possible outcomes makes the decision "unstable." Hence, switching from making a bet to rejecting it, or vice versa, is more a matter of an arbitrary preference than a rational decision.

16.2.10 *Straw man*

Hypothesis testing is introduced and criticized in the section "Bernoulli's swindle and hypothesis testing," starting on p. 67.

Here is my own presentation of frequency-style hypothesis testing. Suppose a new scientific method may have a desirable effect; for example, a new drug may be effective. To simplify things, consider the case of only two hypotheses: H_0 (the "null hypothesis"), meaning that the new method does not work, and H_1 (the "alternative"), meaning that the new method works. On the mathematical side, each of the two hypotheses is represented by a different probability distribution. Data are collected, and the probability is calculated that the observed or more extreme data would be observed if H_0 were true. If the probability is very small (this is often taken to be 5% or 1%; we will use 5%), the null hypothesis is rejected, and the alternative is accepted. In other words, the new scientific method is declared to be effective. The frequentist claim is that this makes the probability of a "false positive" equal to 5%. According to the frequency philosophy, this probability has a frequency interpretation. Namely, in the (sub)sequence of hypothesis tests characterized by the null hypothesis being true, representing the cases when the new scientific method is ineffective, the method will be mistakenly declared to be effective about 5% of the time in the long run.

Diaconis and Skyrms attribute the method of hypothesis testing to Fisher. My description is more in the style of Neyman-Pearson. Diaconis and Skyrms say that "[Fisher] said quite clearly that [hypothesis testing] did not give the probability of effectiveness given the data. [...] You want the probability of effectiveness given the data." This is where the straw man is. I see no reason why all people would desire to know "the probability of effectiveness given the data." I guess Bayesians do. Frequentists may say that they want to minimize the number of false positives, and their method makes the frequency of false positives in the sequence of cases when the null hypothesis is true to be equal to 5%.

On p. 115, the authors say, "... a recent attempt to replicate experimental results, published in leading experimental journals, with p-values less than 0.05, found that less than half of them replicated." I believe that this is an exceedingly unfair and misleading statement. The quoted sentence suggests to people who are not familiar with hypothesis testing that, for some notion of "success," frequentists promised a 95% success rate, but in reality, the method yields less than 50% success rate. As far as I know, frequentists never claimed that the frequency of false positives among all positives would be low. They claim that the frequency of false positives will be low in cases when the null hypothesis is true. Frequentists can be criticized for making useless promises (if this is what Bayesians think) or failing to fulfill their promises, but they should not be criticized for not fulfilling Bayesian aspirations.

The book [6] does not mention an elementary idea that could increase the replication rate. Change the "significance level" in frequency-style hypothesis tests so that the probability of a false positive, given the null hypothesis, is 1%, or even 0.1%. I am not too fond of this idea for the reasons discussed below. Nevertheless, the idea is so simple and obvious that one must wonder why it was not mentioned in [6], even if only to criticize and reject it.

16.2.10.1 *Repairing hypothesis testing*

I will discuss some significant (in my opinion) problems with any attempt to improve hypothesis testing. They apply equally to the traditional frequency-style hypothesis testing, and the modified version discussed on p. 117 of [6]. None of these problems is mentioned in the book under review.

The first problem is that the decrease in the number of false positives will increase the rate of false negatives—this is quite intuitive and supported by theoretical statistics (Neyman-Pearson Lemma). The rate of false negatives is very hard to measure because, typically, negative results are not published. False negatives are not mentioned when the authors discuss methods to improve the replication rate on p. 117. Would the rate of false negatives increase? If yes, how much? Would it matter? It is very hard, perhaps impossible, to determine the (negative) utility of false negatives and false positives. Hence, it is a pure guess as to whether it would be more beneficial to decrease the rate of false positives or false negatives.

The second problem is "social" rather than scientific. Most of the time, scientists publish only positive results for various, rather obvious reasons. Fewer scientists' results will be publishable if research journals tighten their quality standards. Since scientists must get their salaries, tenures, grants and Nobel prizes, they will compensate by performing more experiments. Hence, they will generate more true positives and more false positives. I do not see any obvious reason why the percentage of false positives among all positives would necessarily go down.

The rate at which false positives are generated depends on the rate at which flawed scientific theories are proposed. Likewise, the rate at which false negatives are generated depends on the rate at which solid scientific theories are propounded. The rates at which flawed and solid scientific theories are invented depend on many factors, including the amount of top talent in a given field of science, quality of education, traditions in a specific field, etc. Although some of these factors, such as the quality of education, can be actively improved, I doubt we have enough control. The rates for flawed and solid theories may change due to causes beyond our control.

16.2.11 *Collective responsibility*

On p. 117, the authors propose to improve hypothesis testing. Among other suggestions, we find this: "Prior probabilities (base rates in the field) should not be neglected." Suppose the replication rate in a given field is low. In that case, a given article will have a lower probability of acceptance for publication in a journal than an article of the same quality in a different field where the replication rate is high. In other words, poor scientific practices of other scholars in the same field could hurt a researcher's career.

Criminal trials in the US are very similar to hypothesis tests because the jury has to make one of only two decisions: guilty or not guilty. A jury may also be deadlocked, but this third possible outcome does not affect the discussion below. Statistics show crime rates higher than average for some races in the US. There is strong evidence that membership in some racial groups affected the chances of being convicted by a jury in the past, and this continues, perhaps to a lesser degree. This form of racial discrimination in the justice system is widely considered to be unacceptable.

It seems that it is acceptable that the chances of one's success are affected by the record of other group members in one context (publishing) but not acceptable in a different context (justice system). I do not have an easy philosophical solution to this problem because it involves not only the science of probability but also ethical considerations. The least I can say is that if group membership is deemed irrelevant in one context (justice system) for ethical reasons, we should at least discuss the ethical side of the proposal to consider "base rates" in publishing.

There is also a further consideration. Suppose that statistical analysis shows that the replication rates are lower for a subgroup in a field of science, say, for papers submitted from some country. Should we apply the principle that "Prior probabilities (base rates in the field) should not be neglected" and raise the bar for scientists from that country? The same remark applies to other subgroups of scientists. Should we use base rates for scientists of different genders, ethnicities, and ages?

16.2.12 *Red herring*

The concept of a Dutch book is introduced in Chap. 2 on "Judgment." The examples of Dutch book reasoning are limited to textbook-style artificial examples, predictions markets and arbitrage pricing in financial markets. In prediction markets, people bet on the outcomes of sports events, elections, etc. Arbitrage pricing is a narrow field of finance. There is no hint of how a Dutch book can arise in common situations in everyday life, science, military operations, business, health care, etc.

You can take one of two routes to a movie theater tonight. You believe one is more likely to be jammed than the other. How can you get into a Dutch book situation if you take one of these routes? See also Sec. 6.2.4.

The book [6], like all other books supportive of the Dutch book argument, leaves no doubt that the concept of a Dutch book is a figment of a philosopher's imagination, except for a handful of awkward examples.

16.2.13 *Skepticism*

Chap. 10 of [6] is titled "Induction" and devoted to skepticism. Overall, the discussion is very clear and very detailed. I consider the chapter strikingly incomplete because it ignores reality. By "reality," I mean social reality—I will explain this in detail in a moment. On p. 207, we read, "... it is nevertheless *logically* possible to be a consistent skeptic" (emphasis in original). I agree. Nevertheless, we live in a society that expects us to behave in specific ways to earn rewards such as a salary and avoid punishments such as imprisonment.

Consider religion. One may present many philosophical arguments in favor of or against the existence of God or gods and arguments supporting specific religions. There is a price for any of these philosophical options. This is especially clear in the case of apostasy. If you reject the religion of your community, the community may reject you. Moreover, you can be punished. According to Wikipedia (see [32]), about ten countries have criminal statutes that forbid apostasy or blasphemy as of this writing. Neither apostasy nor its criminal status in some countries add much weight to theological arguments. However, a discussion of religion would be incomplete without an account of actual social practices related to religion.

Privately, scientists can be as skeptical as they wish. An astronomer may believe there is no proof that the Sun is bigger than the Moon. A chemist may believe there are no more than 20 chemical elements. However, in the academic and broader settings of modern society, scientists, engineers, business people, etc., have to adopt certain beliefs and follow specific rules to earn cooperation from other people. In her publications, an astronomer must assume that the Sun is bigger than the Moon, and a chemist must assume that there are more than 20 chemical elements. Otherwise, they would be ostracized.

Statisticians and other people using probability and statistics can be privately skeptical as much as they wish. They can believe that a sequence of coin tosses is exchangeable on Monday, not Wednesday. In modern society's academic and broader settings, they must adopt certain beliefs and follow specific rules. In particular, they must recognize certain sequences as exchangeable and others as not. See Chap. 14 in this book for examples illustrating my point.

My laws (L1)-(L6) codify probability rules enforced in the scientific community. I cannot prove that symmetry or independence exist—I agree with Diaconis' and Skyrms' skepticism. However, (L1)-(L6) are enforced just like the principle of induction is. All scientists, including the most ardent supporters of the Bayesian approach, religiously follow (L1)-(L6). De Finetti's claim that there is no objective probability cannot be disproved. However, there are no real-life examples of the implementation of his philosophy. Subjectivism has the same status as solipsism or relativism—these are all radical, perhaps intriguing, philosophical theories that were never taken seriously in ordinary life. Diaconis and Skyrms may set their own skepticism at any level they wish. The failure to report the level of skepticism universally accepted and enforced in science is a significant deficiency of Chap. 10 of [6].

16.2.14 *Prior problem*

The book [6] mixes science and philosophy. A big problem at the intersection of these two ideologies is caused by terminology. The concept of a "prior" has a fundamentally different meaning in philosophy than in Bayesian statistics. What philosophers call a "prior distribution" is divided into two parts in Bayesian statistics, "prior" and "model." For example, one could say that tosses of a deformed coin form an exchangeable sequence—this is an example of a statistical model. According to de Finetti's theorem, this sequence is a mixture of i.i.d. sequences. The mixing measure is called "prior" or "prior distribution" in statistics. For example, the prior can be the uniform distribution on the interval $[0, 1]$, representing the idea of "no prior information." On the philosophical side, a "prior distribution" refers to the totality of all probabilistic beliefs a statistician holds before collecting the data. To specify all probabilities of all possible events, one needs both the model and the statistical prior. Hence

$$\text{philosophical prior} = \text{statistical prior} + \text{statistical model}.$$

Bayesian statisticians treat statistical priors and models in a completely different way. Statistical models are considered objective (even if nobody makes such an explicit philosophical statement). Statistical priors have a considerably different status. The attitudes towards statistical priors vary widely among Bayesian statisticians. Priors are considered by some to be subjective and by others to be objective. They are often chosen for mathematical convenience.

The book [6] is missing a discussion of the difference between philosophical and statistical priors. The authors discuss pigheaded priors on p. 114. I did not find any discussion of pigheaded models, such as the exchangeable model for Chicago temperatures, presented in Sec. 16.2.2 of this book. The authors do not explain why statistical priors and statistical models merit different philosophical treatment, although their roles are indistinguishable within a philosophical prior.

16.3 "Understanding Uncertainty" by Lindley

16.3.1 *General overview*

Lindley's book is an introduction to probability and statistics that squeezes an enormous amount of material into a short volume (less than 250 pages) at a relatively relaxed pace. The reader will learn about confounding variables, the Bayes rule and hypothesis testing, and the basic concepts of dependence, expectation and variance. If the philosophical remarks were removed, the book would make an excellent undergraduate statistics textbook.

Curiously, Lindley included Sec. 2.9, which reviews mathematics at the elementary school level. For example, Lindley reminds the reader that if $a + b = c$, then $a = c - b$. I wonder how many readers who needed a tutorial at this level could absorb non-trivial ideas in the remaining part of the book.

16.3.2 *There is no objective probability*

Lindley's book [7] is a striking example of intellectual hypocrisy. Lindley repeatedly makes two contradictory claims:

(i) There is no objective probability.
(ii) Good (rational, beneficial) ways of linking probability values with reality exist.

The first and second claims are presented in fundamentally different ways. The first claim is stated repeatedly, explicitly and conspicuously. The second claim is implicit in the book, but it is smuggled very effectively—it exploits reader's learning instinct based on examples. Below is a collection of quotes from [7] in which Lindley pronounces strict subjectivity of probability.

Page 1: "[...] uncertainty is a personal matter; it is not *the* uncertainty but *your* uncertainty. [...] Statements of uncertainty are personalistic, they belong to the person making them and express a relationship between that person and the real world about which a statement is being made. In particular, they are not objective in the sense that they express a property that is the same for all of us." (emphasis in original)

Page 12: "[...] probability does not exist as a property of the world in the way that distance does, for distance between two points, properly measured, is the same for all of us, it is objective, whereas probability depends on the person looking at the world, on you, as well as on the event, that aspect of the world under consideration."

Page 91: "This is a convenient point to remind the reader that there is almost nothing in this book to say what your beliefs should be; only how they should fit together, or cohere. Cromwell's rule is a slight exception, but all it does is to exclude values 0 and 1 in most circumstances, because their use can lead to what many people consider unsatisfactory results."

Page 101: "Notice that there is no obligation on you to have the same beliefs in the two outcomes, only that if you do, your probabilities are both one half."

Page 236: "Before leaving the discussion, let me emphasize the point made at the beginning, namely that the emphasis is on the methodology of the discussion and not on its economic soundness. It would be possible for the reader, acting as another 'you', to replace some, if not all, of the probabilities used above, by other probabilities in order to produce an argument that supports the trickle-down effect. The discussion in this section, indeed throughout the book, is not intended to be partisan but only to demonstrate a form of reasoning, using uncertainty, intended to shed new light on a problem."

Page 236: "But there is no obligation on you to accept these beliefs, so that you can believe that, when you have tossed the coin 10 times with heads every time, the next toss will probably be tails, to make up the deficit; or a pregnant woman believes that the child she bears is male. Neither of these beliefs is wrong; the most that can be said is that they are unusual or incoherent."

Page 242: "The thesis is personal [this sentence was displayed for emphasis]. That is to say, it is a method for 'you'. [...] Our theory only admits one probability and one utility. When cooperation is present, as in the jury room or in the board room, a "you" is not unreasonable, but not with conflict. [...] Conflict is the most important mathematical and social problem of the present time."

The emphasis on conflict illustrates Lindley's belief that there is no objective probability because otherwise, the obvious resolution of the conflict between incompatible subjective probabilistic opinions would be to find the objective probability.

Lindley prefers to call probability "personal" rather than "subjective;" he often uses the words "you" and "your" for the same purpose. Lindley ignores many situations when probability assignments are subcontracted. Companies, government agencies and the military hire statisticians or external consulting companies to analyze data. If a pharmaceutical company releases a statement saying that the probability of side effects for a given drug is 7%, whose "personal" probability is it?

16.3.3 *Objective probability—indoctrination via examples*

Lindley effectively indoctrinates his readers that there is an objective probability by presenting only these examples of probabilities that are widely accepted in science. There is one insignificant exception that will be discussed below. I find Lindley's protestations that he does not believe in objective probability impossible.

16.3.3.1 *Objective bias*

Here is a long list of examples of the standard or popular probability models and distributions in [7].

Page 3: "the chance of an ace is 4 divided by 52, or 1 in 13."

Page 33: "In some cases you might feel that every ball, and therefore every number, had the same chance of being drawn as every other; that is, all numbers from 1 to 100 had the same uncertainty."

Page 37: "Suppose you meet a stranger. Take the event that they were born on March 4 in some year. You are uncertain about this event but the comparison with the urn is easy and most of you would announce a probability of 1/365, ignoring leap years and any minor variations in the birth rate during the year."

Page 42: "There is only one chance in about 14 million that your ticket will win." This refers to a lottery.

Page 55: "The electricity grid, carrying supplies around the country, is designed so that a failure in one part of it can be compensated by re-routing the electricity[...]. [...] In the original calculations about supply, failures were supposed independent and the calculations of the small probability of a place having no supply were based on this assumption. However, one cause of failure is the trees fouling the electricity cables, especially in a high wind in the spring when growth of the trees is vigorous. Thus a storm in April can cause several failures, revealing that independence of interruption events is not a sensible assumption."

Page 70: "[...] a patient who truly has cancer has only one chance in 100 of slipping past the test undetected and therefore is almost certain, probability 0.99, of being detected [...]. [...] a cancer-free patient still has probability 0.05 of a positive test result."

Page 103: "Now let the pin be tossed a number of times under conditions that remain stable; you do not, for instance, alter the tossing procedure. To be specific, let the results of 10 tosses in order be $UUDUDUUUDD$ and denote this result by x. Notice that six times the pin fell uppermost and four times it fell with the point down, giving a frequency of Us of 0.6. You are about to toss an eleventh time and are uncertain about the event U on that occasion. What is your probability [...]? A natural response is 0.6, the frequency in the series of 10 tosses[...]."

Page 104: "The key point is that the order of the Us and Ds may matter with weather, but usually not with drawing pins."

Page 104: "Suppose that when you think about the possible results of the 10 tosses, you feel that your probability for any series depends only on the number of times the pin falls with the point upward and not on the arrangement of the Us and Ds in the series."

Page 110: "[...] your probability that the coin is fair, and being tossed fairly, is 0.92, but admit that other values are possible with small probabilities."

Page 116: "Those who accept quantum mechanics accept exchangeability as part of that acceptance and therefore have chances."

Page 116: "We now have two methods of assessing probabilities: using the concept of cases, which have equal uncertainty, the classical method, and that based on frequency allied with the concept of exchangeability. The former only applies to a limited class of situations, like games that use cards or dice. The second is of such wide use that probability is often confused with frequency."

Page 124: "[...] all tosses of a coin could fall heads, but it is unlikely [...]."

Page 125: "[...] you would ordinarily make some assumption of exchangeability of yourself with the patients who took part in the trial."

Page 129: "[...] the unemployment rate among the Blacks is 75% (48 out of 64), whereas among the Whites it is only 12.5% (8 out of 64). A black person is six times more likely to be out of work than a white person." Here, probability ("likely") is quite explicitly identified with frequency ("75%" and "12.5%").

Page 132: "The main lesson to be learnt from the material in this chapter is that the relationship between two uncertain events is not always what it appears to be because there may be a third event that influences them both and distorts the apparent connection." If there is no objective probability, how can a lesson be learnt? A lesson about what? Lindley is not talking here about coherence.

Page 226: "for elections are not usually judged exchangeable."

In addition, Lindley discussed popular probability distributions: binomial (p. 135), normal (p. 149), and Poisson (p. 139). Chap. 12 is devoted to a detailed analysis of eight examples. Each example uses standard (popular) probability assignments.

16.3.3.2 *Non-standard opinions*

It is instructive to review the only explicit example (that I was able to find in the book) of probabilities that are not mainstream (p. 236):

> "[...] you can believe that, when you have tossed the coin 10 times with heads every time, the next toss will probably be tails, to make up the deficit; or a pregnant woman believes that the child she bears is male. Neither of these beliefs is wrong; the most that can be said is that they are unusual or incoherent."

It is incredible how misleading a short and seemingly straightforward paragraph can be.

First, the claim that "neither of these beliefs is wrong" is undermined by a suggestion that they might be "incoherent." This is significantly misleading because every probability belief may be incoherent. If you believe that the probability of an event is p, you may be incoherent by believing that the probability that the event will not happen is different from $1 - p$. There is an implicit suggestion in Lindley's remark that unusual probabilistic beliefs are more likely to be incoherent than mainstream probabilistic beliefs. While this may be true if you choose a completely arbitrary set of numbers for your probabilities, I doubt that many people arbitrarily choose probability values. One of the main problems with probability assessment is its dependence on emotions, but this does not mean that probabilities skewed by emotions are incoherent. Both examples given by Lindley illustrate my point. Some gamblers feel a transcendental balance exists in the universe; therefore, a long run of heads must be balanced by tails, or at least by an increased chance of tails. In some cultures, sons are valued more than daughters, so a pregnant woman may hope for a "better" pregnancy outcome.

Let me show how both "unusual" claims can be parts of coherent belief systems. Every time a coin is tossed, count the number of heads among the six most recent results. Call this number k. Then let the probability of heads on the next toss be $p = 1 - (k+1)/8$. Note that p is never zero or one—I follow here Lindley's advice, called "Cromwell's rule," to avoid assigning values 0 or 1 to unknown events. My model is clearly coherent. It is inspired by "mean reverting" models. As for the pregnant woman, she can consider the case of her baby to be independent of all other pregnancies. This is coherent no matter whether there are any "objective" reasons for why she is different from other women. In other words, it is unnecessary to base exchangeability (or its lack) on empirical observations to achieve coherence.

The two examples given by Lindley are misconceptions with little practical significance. Repeated coin tossing is not a popular activity, but it is probabilistically equivalent to betting on red or black at a casino. Actually, the chance of red (or black) is slightly lower than 1/2; it is 18/37 or 18/38, depending on the casino, but this is irrelevant for this discussion. No matter what beliefs you have concerning the dependence of the probability of red on the past results, hence, no matter what betting strategy you choose, your long-run loss distribution will be the same. So, unusual beliefs about red and black outcomes are irrelevant from a practical point of view. They will not disadvantage the holders of unusual opinions relative to the gamblers who hold scientifically accepted probabilistic opinions.

There is anecdotal evidence that some, perhaps many, expectant women have strong opinions about the sex of their future child. The practical significance of such beliefs depends on the culture. In some cultures, baby clothing is color-coded (blue for boys and pink for girls). So it is possible that the prospective parents could buy the incorrectly-colored baby clothes before the child's birth due to the extreme probabilistic beliefs about the child's gender. This mistake will likely be made only by middle-class (or wealthier) parents. They can easily absorb the loss. I cannot think of any other practical consequence of strong beliefs about a child's gender.

Lindley does not mention well-known cases when overconfidence can bring disastrous results, even death. Some people believe that they can handle drinking and driving. They believe that the probability of killing someone in an accident is very low, even if they are drunk. Some other people keep loaded guns at home at locations accessible to their children. They believe that the probability that a child could find the gun (by chance) and hurt oneself or someone else is negligibly small. I will never know whether Lindley would repeat here his comment, "Neither of these beliefs is wrong; the most that can be said is that they are unusual or incoherent." If he did, he would expose subjective philosophy's real dangers. If he did not, then he would implicitly admit that the subjectivist extremism is unsustainable in real life.

16.3.3.3 *Social pressure*

Recall the quote from Sec. 16.3.3.2. Lindley says, "Neither of these beliefs is wrong." The claim ignores the social pressure. Every person can believe that $2 + 2 = 7$ and that the Moon is made of green cheese if she chooses to do so. Nevertheless, society puts enormous ideological pressure on its individuals. To function normally in society, they often have to suppress their beliefs. Take, for example, the suppression of some religious practices in intolerant societies. An account of probability theory may contain arbitrarily crazy philosophical ideas. However, it is honest only if it accurately describes the probabilistic beliefs enforced by social pressure. Statisticians and other scientists who want to work in academia or the corporate world must use mainstream probabilistic methods. Textbooks on statistics contain the current canon of probabilistic beliefs in the form of mathematical models. To ignore them is tantamount to rejecting opportunities to work at universities, many companies and many professions. I am tempted to accuse Lindley of hurting people by giving them harmful advice, but nobody takes the claim that "Neither of these beliefs is wrong" seriously, so no harm is done.

16.3.3.4 *Implicit objectivity*

I find the following remark on p. 2 of [7] puzzling. "You are uncertain about some aspect of the world and that uncertainty does not refer solely to you, or solely to the world, but describes a relationship between you and that world." We can replace "uncertainty" in this remark with any subjective concept, say, "beauty of colors," and the remark becomes (or stays) obvious. Any choice of the most beautiful color "does not refer solely to you, or solely to the world, but describes a relationship between you and that world." Either Lindley's remark is obvious and insignificant, or perhaps he was subconsciously hinting at some objectivity in probability.

16.3.3.5 *Mirage of objectivity*

I was struck by the following declaration made by Lindley on p. 78: "The lesson to be learned from this study of weather forecasts applies generally and warns us to be careful in the specification of the uncertain events that are being referred to. This also applies to your knowledge base. It is not merely a question of calculating with probabilities but also one of relating the ingredients of your probability statements to reality."

Didn't Lindley admit that probability is at least partly objective? Alas, no. The careful examination of the context shows that all that Lindley meant was that the family of all events should be precisely specified, and probabilities assigned to these events should be coherent.

16.3.3.6 *Learning*

On p. 81 of [7] we find "For any statement or event, F, you have a probability $p(F)$ on some knowledge base. Next, suppose that you acquire new evidence, E, that bears on the truth of F, so affecting your uncertainty of F; this will change your probability to $p(F \mid E)$ and learning is accomplished by your change from $p(F)$ to $p(F \mid E)$." What is "learning"? I would say that it is a process in which our knowledge is replaced with better knowledge. New knowledge can be better because it is more extensive than the old knowledge or more accurate than the old knowledge. "Learning" never indicates a mere change in knowledge; it always implies improvement. Hence, Lindley indicates that probability assessments can be improved. This contradicts his claim that none of the coherent probability views of the world is better than any other.

16.3.3.7 *Massive agreement*

On p. 115, Lindley points out that "Experience shows that there is a massive agreement about some series being exchangeable, so that the objectivity can be at least a convenient approximation."

There are many massive agreements. There is a massive agreement that Venice is a beautiful city. There is a massive agreement in many countries on a specific religion. There is a massive agreement in Western countries that white is the best color for a bride's gown. Lindley mistakes "objectivity" for "cooperation." When there is a massive agreement, a large group may cooperate on a major project. For example, since there is a massive agreement that Venice is a beautiful city, Italy's government may enjoy the approval of the Italian voters for spending massive amounts of money on saving Venice from flooding. Nothing can prove that Venice is objectively beautiful.

16.3.3.8 *Inevitable agreement*

On p. 194, Lindley states that "It is only with the accumulation of more data that agreement between investigators can be attained and the theory given a probability near to 0 or 1, so that, in the latter case, it can be reasonably asserted to be true, whereas in the former, it can be dismissed."

The claim is repeated in a powerful form on pp. 236–237:

> But there is no obligation on you to accept these beliefs, so that you can believe that, when you have tossed the coin 10 times with heads every time, the next toss will probably be tails, to make up the deficit; or a pregnant woman believes that the child she bears is male. Neither of these beliefs is wrong; the most that can be said is that they are unusual or incoherent. At first sight, this extremely liberal view that you can believe what you like looks set to lead to chaos in society, with us all having different opinions and acting in contrary ways. However, coherence mitigates against this. We saw in the simple example in Sec. 6.9 of the red and white urns that, whatever your initial belief about the color of the urn, provided you updated this belief by Bayes rule, the continual withdrawal of more white balls than red would raise your probability that the urn was white to nearly 1, so that everyone would be in the same position, irrespective of any initial disagreements.

Lindley's claims are based on several implicit essential assumptions:

(i) Each of the probability models chosen by different investigators leads to a sequence of probabilities converging to a limit.
(ii) The limit is the same for each of the models.
(iii) The amount of data collected in practice is sufficient to bring the converging probabilities close to each other in real time, not only in the mathematical limit when the amount of data is infinite.

All three assumptions are realistic but based on the interpretation of probability as an objective quantity. The first two assumptions often result from the exchangeability assumption in statistical practice. The third assumption applies when the prior distributions are not "dogmatic" (see Sec. 16.2.4.2).

Lindley mentioned "chaos in society" but did not discuss any real examples of such "chaos." Religion is a clear example of opinions that do not tend to converge. One could propose that religious beliefs are not coherent, but I do not believe this explains religious diversity well. It is more likely that religious people do not use exchangeability in situations where scientists would and that the priors used by religious people are so concentrated that no amount of data can override them. So, yes, there is "chaos in society," proving that coherence alone is useless.

Coherence is an incredibly weak assumption. One cannot derive exchangeability or non-dogmatic priors from coherence. Lindley's remark that "coherence mitigates against this" is plainly false.

16.3.4 *Why should anyone care about probability?*

On pp. 74–75, Lindley says

> The upshot of the material in the last four sections is that we have five, rather distinct methods of establishing the rules of probability. These are
>
> 1. Comparison with a standard.
> 2. Avoidance of Dutch books in gambling.
> 3. The use of scoring rules.
> 4. An extension of logic.
> 5. Action in the face of uncertainty.

This list is an example of a "category mistake" in which apples and oranges are thrown into one bag, and one pretends they are all alike.

The "rules" 2 and 3 are particular cases of 5. Gambling and making guesses rewarded by scoring are actions in the face of uncertainty.

Rules 1 and 4 may be intellectual choices of some people, but they do not address the benefits of probability. I can compare my probabilities with a standard, but what do I gain by doing so? Very few people care whether probability is an extension of logic. And why should they?

Rules 2 and 3 address the question of the benefits of using probability. Lindley explains in his book how probability can be applied to eliminate Dutch books. Lindley earns my highest praise for adding "in gambling" to rule 2. He is realistic about the incredibly narrow field of applications of the Dutch book argument.

The scoring rule idea is useless without objective probability. On p. 77, Lindley writes. "It was seen, with the discussion of scoring rules in Sec. 5.8, that when verification is possible a penalty score can be constructed, which will provide assistance in assessing the meteorologist's ability, not only on one occasion, but over a period. This has been used with weather forecasts of rain at a specified place in the United States, where the professionals performed well, achieving a low penalty score." The success is admirable, but the knowledge of the success is useless to other people if the method of achieving low penalty scores cannot be described and copied by other probability assessors. If the method can be described, it is a form of objective probability. Objective science is a system of communication that passes successful methods from their inventors to other scientists and the general public.

People can test their scoring abilities online at [97]. It appears that some scorers are more successful than others. I did not find anything in [7] that would explain some scorers' excellent performance compared to others' poor performance.

On p. 239 of [7] we find "If you have a situation in which uncertainty plays a role, then probability is the tool you have to use. To use anything else will lay you open to the possibility of your being shown up to be absurd, say in the sense of a Dutch book [...] being made against you." If the danger of finding oneself in a Dutch book situation is the main (and only) reason for using probability, then we can safely abandon this science. Dutch books occur only in books on the philosophy of probability.

16.3.5 *Racism*

Sec. 8.7 of [7] is devoted to "Spurious association." It presents real data (slightly modified to simplify them) on race (Blacks and Whites) and crime rates. The data show a higher rate of a specific crime among Blacks (41%) than Whites (17%). Then Lindley reports that the unemployment rates for the two ethnic groups are dramatically different: 75% for Blacks and 12.5% for Whites. Lindley concludes that "Here we have one factor, race, being blamed for crime, when the culprit is really unemployment." It is abundantly clear that Lindley promotes objective thinking. The "spurious association" is presented as an objective fact that should affect our personal probabilities.

Elsewhere in the book, Lindley cautions against identifying frequency and probability (he gives three reasons against the identification on pp. 103–104) and supports "unusual" probabilistic opinions (see Sec. 16.3.3.2). Both of these intellectual choices can be used to justify an opinion that Blacks are more likely to be criminals than Whites. Lindley does not comment on the rationality of stubborn but coherent racism.

By the way, Lindley did not say whether the data used in Sec. 8.7 of [7] represented the whole United Kingdom or a part of it. According to the UK government figures published online at [98], the unemployment rate never exceeded 20% for Blacks and never exceeded 10% for Whites between 2004 and 2019.

Lindley makes the following statement on p. 130: "Two events are independent if, whatever you learn about one, does not affect your uncertainty about the other." This can be used to approve prejudices, for example, against religious or ethnic groups. Given the claim that no probability belief is wrong unless incoherent (see Sec. 16.3.3.2), Lindley gives a license to ignore all evidence that could invalidate a prejudice.

16.3.6 *Hypothesis testing*

It is interesting to see what Lindley says about hypothesis testing (see Sec. 16.2.4.3), given the disappointment of Diaconis and

Skyrms with the frequency-style hypothesis testing. On p. 203, Lindley writes: "In words, the equation expresses your opinion, in the form of odds of the null hypothesis on the basis of the results of the trial, appearing on the left, in terms of the same odds before the trial, on the right. The latter, multiplied by the likelihood ratio, expresses how the probability of the data on the null hypothesis differs from that on the alternative."

Recall that all probabilities are personal and subjective, according to Lindley. So "odds before the trial" can only represent the beliefs of the experimenter; they do not represent any objective quantity. It follows that the "odds of the null hypothesis on the basis of the results of the trial" depend, possibly significantly, on the personal opinion held by the investigator before the trials.

It is only natural to expect that the inventor of a new drug would hold a strong positive opinion about the drug's effectiveness. This enthusiastic opinion is not wrong because no probabilistic opinion can be wrong, according to Lindley. Thus, the decision to approve a new drug may be based to an arbitrarily large extent on the personal opinion of the inventor. One has to wonder why this "Bayesian hypothesis testing" is supposed to be better than the traditional frequency-style hypothesis testing unless some authority imposes rules governing the "odds before the trial." Then, one has to ask who that authority might be and how the rules might be chosen.

16.3.7 *Probability and chance*

Subjectivists keep trying to perform an impossible balancing act. On the one hand, they claim that all probability is subjective (or personal). On the other hand, they must account for all the existing scientific evidence for objective probability so that they do not appear totally irrational. On p. 115 of his book, Lindley analyzed repeated tosses of a thumbtack as follows,

> Another feature of the Bernoulli θ is that it has a degree of objectivity in the sense that if Peter and Mary both judge a series to be exchangeable, then the value of θ, as a limiting frequency, will be common to them both, though unknown to them both. The objectivity is limited though because if Paul disagrees with exchangeability, θ may not have a meaning for him. Experience shows that there is a massive agreement about some series being exchangeable, so that the objectivity can be at least a convenient approximation.
>
> The upshot of these considerations is that θ, while it obeys the rules of the probability calculus, is not a probability in the sense of a belief. As a result, we prefer to give it a different name and it is often referred to as a chance.

Recall that according to de Finetti's theorem, the sequence of results, if it is exchangeable, can be represented as a mixture of i.i.d. sequences with the mixing measure identified with the distribution of θ.

Lindley wrote on p. 91, "This is a convenient point to remind the reader that there is almost nothing in this book to say what your beliefs should be; only how they should fit together, or cohere." If we adopt this philosophical position, Peter and Mary's coherent opinions may be close to each other or not, even if they are perfectly rational. It follows that the claim "... if Peter and Mary both judge a series to be exchangeable, then the value of θ, as a limiting frequency, will be common to them both ..." is no more than a mathematical tautology, true irrespective of any objective properties of the sequence.

The word "judge" in the phrase "... Peter and Mary both judge a series to be exchangeable ..." strongly indicates that Lindley believed in objective symmetries related to probability. If there are no objective symmetries related to probability, then there is nothing to "judge." In a world without objective symmetries related to probability, Peter and Mary could only arbitrarily "declare" a series to be exchangeable.

16.3.8 *Conflict resolution*

The following quote comes from the very end of [7] (pp. 241–242):

> Any sound appreciation of the thesis presented in this book must recognize a severe limitation of that thesis [...]
>
> The thesis is personal.
>
> [centered in original]
>
> That is to say, it is a method for "you". [...] Our theory only admits one probability and one utility. When cooperation is present [...] a "you" is not unreasonable, but not with conflict. [...] My hope is that today, somewhere in the world, there is a young person with sufficient skill and enthusiasm to be given at least five years to spend half their time thinking about decision making under conflict. They will need to be a person with considerable skill in mathematics, for only a mathematician has enough skill in reasoning and abstraction to capture what hopefully is out there, waiting to be discovered. Conflict is the most important mathematical and social problem of the present time.

The above text is followed by only one short paragraph on an unrelated topic. Delegating these ideas to the very end of the book indicates that Lindley attached great importance to conflict resolution.

The word "conflict" has different meanings. Lindley might have interpreted it as any situation that is not necessarily antagonistic when two people have different probabilistic opinions. But his dramatic appeal suggests that he understood "conflict" as not only a difference in probabilistic opinions but also a conflict of interests.

The civil law goes back to antiquity, most notably to the Roman Empire. The system of civil law is partly a system for conflict resolution. Another old and well-known method of conflict resolution is consensual compromise via negotiations. Various ideologies include ideas on conflict resolution. An example of such an idea comes from the utilitarian philosophy: "The good of a society is the sum of happiness of the individuals in that society." This may lead to the "optimal" solution of conflicts. Alternatively, one could remove the source of the conflict. Religious suggestions along these lines include "If someone slaps you on one cheek, turn to them the other also" (Bible) and "Desire is the root of evil" (Buddha). Because of the long history of philosophical research on various ideas related to conflict resolution, Lindley's suggestion that a mathematician could make substantial contributions to this field after a five-year effort is truly off the wall.

There is an impractical but simple and effective method of solving conflicts between subjectivist holders of different probabilistic opinions. One of the coherent probabilistic views of the world should be chosen arbitrarily (say, democratically or by lottery), and all people involved in the conflict should adopt this coherent family of opinions. According to Lindley, since objective probability does not exist, nobody will have to give up a rational opinion. One would have to check whether the adoption of a new coherent set of probabilities would generate a Dutch book against anyone (since some decisions might have been made in the past) but in most practical situations, it would not (see Sec. 6.2.4).

Lindley's intentions are commendable, but he misses the fundamental solution to his quandary. The incredibly powerful and successful solution to conflict resolution is objective science. People of different persuasions will agree on science if they agree on anything. Consider an example of enemy states who signed a treaty limiting the number of tanks on each side to 5,000. Each side may try to cheat by hiding tanks in the woods, calling tanks "armored vehicles," etc. But if one state documents the presence of 4,000 tanks in one part of the other state's territory and 3,000 tanks in some other part, the offending state will not claim that $4,000 + 3,000 = 5,000$. Why not if all other cheating methods seem acceptable or at least practiced? The laws of science are so fundamental that if you deny them, no trust between the parties is left. Saying that $4,000 + 3,000 = 5,000$ effectively removes any basis for cooperation or negotiation. If the laws of science are ignored, nothing is left except naked force.

In the last paragraph of his book, Lindley writes "People often accuse me of putting the Bayesian argument forward as if it was a religion." Then he denies that he is a religious subjectivist and claims to be rational. However, only religious zeal can blind someone to the fact that the science of statistics is a collective attempt to find as much objective probability in uncertain situations as possible. Bayesian statisticians are no less objective than the frequentists. The best-known and widely practiced method of finding objective values of probabilities is based on relative frequencies in long sequences of independent and identical trials. Lindley dismisses the method by calling it "chance" (to distinguish it from probability) and devotes a small section of his book to it. I agree with Lindley that it is a good idea to have reliable, practical methods for conflict resolution under uncertainty. They are called objective probability. I hope that no mathematician wastes five years searching for something already well-known.

By the way, I would call Lindley a "fanatical" subjectivist, not a "religious" subjectivist. Not all religious people are fanatical, and not all fanatical people are religious.

16.3.8.1 *Ethics of helpful advice*

The flip side of Lindley's conflict resolution suggestion is that the personal probability theory creates problems in cooperative situations as significant as those in conflict situations.

According to the subjective philosophy of probability, there is no objective probability, so no member of a cooperative group of people is closer to the truth than any other. Hence, there is no rational reason to think that offering others your opinion on probability will bring them closer to the truth. If they are consistent in their own way, your probabilistic opinion will not eliminate any Dutch books because these have already been eliminated.

Reporting observed facts to other group members is usually considered helpful, but I do not see why it would be helpful in the context of subjective probability. Learning new facts may affect probabilistic opinions via conditioning, but how would it eliminate any Dutch books if there are none to start with?

16.3.8.2 *Enforcement of subjective opinions*

Enforcing subjective opinions, such as those about beauty, is considered unethical because such enforcement limits freedom for no good reason.

I believe that it is ethical to enforce subjective probabilistic opinions. My argument may apply to non-probabilistic subjective opinions—I will let the reader decide.

People have subjective opinions about drinking and driving—some consider this a hazardous habit, and some do not. In probabilistic terms, the question is about to what extent drinking increases the probability of an accident. Drivers have no choice with whom to share the road. Hence, the law against drinking and driving addresses the anxieties of many drivers and provides an intangible but real and objective benefit to them. The law is justified by the principle of maximizing the sum of utilities of all people, even if objective probability does not exist.

The same argument does not apply to many safety regulations; for example, building codes. These regulate the form and function of many elements of houses, such as the frame, egresses, electrical wiring, etc. Without objective probability, one cannot decrease the probability of dying in a house that collapses in an earthquake or catches a fire. By law, the owner cannot exercise his subjective opinions about these probabilities. Building codes implicitly assume that probability is objective.

Chapter 17

Utopia

I end the book by describing my dream. I hope that, in the not-too-distant future, undergraduate textbooks will present Kolmogorov's axioms as the mathematical basis of probability and (L1)-(L6) as the scientific laws of probability. I hope discussing probability's "true nature" will be relegated to philosophical monographs. In other words, I hope probability will be taught just like Newton's laws of motion.

I hope that frequency and Bayesian statisticians will recognize that the laws (L1)-(L6) are the foundations of statistics.

I hope the subjective philosophical theory of probability will be abandoned (the frequency philosophy was abandoned long ago), and Cournot's principle will be restored to its full glory.

Bibliography

[1] K. Burdzy, *The search for certainty. On the clash of science and philosophy of probability.* World Scientific Publishing Co. Pte. Ltd., Hackensack, NJ (2009).
[2] K. Burdzy, *Resonance—from probability to epistemology and back.* Imperial College Press, London (2016), ISBN 978-1-78326-920-4.
[3] R. von Mises, *Probability, statistics and truth*, English edn. Dover Publications, Inc., New York (1981).
[4] B. de Finetti, *Theory of probability: a critical introductory treatment. Vol. 1.* John Wiley & Sons, London-New York-Sydney (1974), translated by Antonio Machì and Adrian Smith, With a foreword by D. V. Lindley, Wiley Series in Probability and Mathematical Statistics.
[5] Wikipedia, Five Ws, (2021), retrieved: 19 May 2021, `https://en.wikipedia.org/wiki/Five_Ws`.
[6] P. Diaconis and B. Skyrms, *Ten great ideas about chance.* Princeton University Press, Princeton, NJ (2018), ISBN 978-0-691-17416-7.
[7] D. V. Lindley, *Understanding uncertainty*, revised edn., Wiley Series in Probability and Statistics. John Wiley & Sons, Inc., Hoboken, NJ (2014).
[8] M. Galavotti, *Philosophical introduction to probability.* CSLI Publications, Stanford, Calif. (2005).
[9] D. Gillies, *Philosophical Theories of Probability.* Routledge, London (2000).
[10] R. Weatherford, *Philosophical foundations of probability theory.* Routledge & K. Paul, London (1982).
[11] T. Fine, *Theories of Probability. An Examination of Foundations.* Academic Press, New York (1973).
[12] J. Keynes, *A Treatise on Probability.* MacMillan, London (1921).
[13] R. Carnap, *Logical Foundations of Probability.* The University of Chicago Press, Chicago, Ill. (1950).
[14] K. Popper, *The Logic of Scientific Discovery.* Harper and Row, New York (1968).
[15] H. Primas, Basic elements and problems of probability theory, *Journal of Scientific Exploration* **13**, pp. 579–613 (1999).
[16] M. DeGroot, *Optimal Statistical Decisions.* McGraw-Hill, New York (1970).

[17] P. C. Fishburn, *Utility theory for decision making*, Publications in Operations Research, No. 18. John Wiley & Sons, Inc., New York-London-Sydney (1970).
[18] L. J. Savage, *The foundations of statistics*, revised edn. Dover Publications, Inc., New York (1972).
[19] R. Royall, *Statistical Evidence: A Likelihood Paradigm.* Chapman & Hall, London (1997).
[20] P. Bandyopadhyay, G. Brittan and M. Taper, *Belief, Evidence, and Uncertainty: Problems of Epistemic Inference.* Springer, Cham (2016).
[21] M. Taper, G. Brittan and P. Bandyopadhyay, Statistical inference and the plethora of probability paradigms: A principled pluralism, (2019), philArchive; `https://philarchive.org/rec/TAPSIA`.
[22] A. Hájek, Interpretations of probability, in E. N. Zalta (ed.), *The Stanford Encyclopedia of Philosophy*, winter 2012 edn. Metaphysics Research Lab, Stanford University (2012), `https://plato.stanford.edu/archives/win2012/entries/probability-interpret/`.
[23] D. Mellor, *Probability: a philosophical introduction.* Routledge, London; New York (2005).
[24] I. Hacking, *An introduction to probability and inductive logic.* Cambridge University Press, Cambridge, U.K.; New York (2001).
[25] J. von Plato, *Creating Modern Probability.* Cambridge Univ. Press, Cambridge (1994).
[26] M. Heller, *Philosophy of Chance. A cosmic fugue with a prelude and a coda.* Copernicus Center Press, Kraków (2012).
[27] J. Pitman, *Probability.* Springer-Verlag, New York (1993).
[28] S. Ross, *A First Course in Probability.* Pearson Prentice Hall, Upper Saddle River (2006), 7th ed.
[29] D. F. Anderson, T. Seppäläinen and B. Valkó, *Introduction to probability*, Cambridge Mathematical Textbooks. Cambridge University Press, Cambridge (2018).
[30] M. Baghramian and J. A. Carter, Relativism, in E. N. Zalta (ed.), *The Stanford Encyclopedia of Philosophy*, Spring 2022 edn. Metaphysics Research Lab, Stanford University (2022), `https://plato.stanford.edu/archives/spr2022/entries/relativism/`.
[31] P. Feyerabend, *Against method: outline of an anarchistic theory of knowledge.* NLB, London; Humanities Press, Atlantic Highlands (1975).
[32] Wikipedia, Apostasy, (2021), retrieved: 13 May 2021, `https://en.wikipedia.org/wiki/Apostasy`.
[33] J. Derrida and L. Venuti, What is a "relevant" translation? *Critical Inquiry* **27**, pp. 174–200 (2001).
[34] G. Marsaglia, The Marsaglia random number CDROM: including the diehard battery of tests of randomness, (1995), Florida State University.
[35] D. Ruelle, *Chance and Chaos.* Princeton University Press, Princeton, N.J. (1991).
[36] Wikipedia, Brownian motion, (2021a), retrieved: 8 May 2021, `https://en.wikipedia.org/wiki/Brownian_motion`.

[37] Wikipedia, Brownian model of financial markets, (2021b), retrieved: 8 May 2021, `https://en.wikipedia.org/wiki/Brownian_model_of_financial_markets`.
[38] H. Kyburg and H. Smokler (eds.), *Studies in Subjective Probability.* Wiley, New York (1980), 2nd ed.
[39] B. de Finetti, *Theory of probability: a critical introductory treatment. Vol. 2.* John Wiley & Sons, London-New York-Sydney (1975), translated from the Italian by Antonio Machì and Adrian Smith, Wiley Series in Probability and Mathematical Statistics.
[40] Wikipedia, Immaculate conception, (2023), retrieved: 30 January 2023, `https://en.wikipedia.org/wiki/Immaculate_Conception`.
[41] D. Wallace, *The Emergent Multiverse: Quantum Theory according to the Everett Interpretation.* Oxford University Press, Oxford (2014).
[42] H. Kyburg, Subjective probability: criticisms, reflections, and problems, *Journal of Philosophical Logic*, pp. 157–180 (1978).
[43] J. O. Berger, *Statistical decision theory and Bayesian analysis*, 2nd edn., Springer Series in Statistics. Springer-Verlag, New York (1985).
[44] S. J. Press and J. M. Tanur, *The subjectivity of scientists and the Bayesian approach*, Wiley Series in Probability and Statistics: Texts, References, and Pocketbooks Section. Wiley-Interscience [John Wiley & Sons], New York (2001).
[45] Wikipedia, Five ways (Aquinas), (2020), retrieved: 19 May 2020, `https://en.wikipedia.org/wiki/Five_Ways_(Aquinas)`.
[46] Wikipedia, Pascal's wager, (2021), retrieved: 10 June 2021, `https://en.wikipedia.org/wiki/Pascal%27s_wager`.
[47] M. Tegmark, *Our Mathematical Universe: My Quest for the Ultimate Nature of Reality.* Knopf, New York (2014).
[48] Wikipedia, Boltzmann brain, (2021a), retrieved: 10 June 2021, `https://en.wikipedia.org/wiki/Boltzmann_brain`.
[49] Wikipedia, Inattentional blindness, (2021b), retrieved: 24 May 2021, `https://en.wikipedia.org/wiki/Inattentional_blindness`.
[50] Wikipedia, Shakespeare authorship question, (2021c), retrieved: 24 May 2021, `https://en.wikipedia.org/wiki/Shakespeare_authorship_question`.
[51] Wikipedia, Phlogiston theory, (2021d), retrieved: 8 June 2021, `https://en.wikipedia.org/wiki/Phlogiston_theory`.
[52] Wikipedia, Luminiferous aether, (2021e), retrieved: 8 June 2021, `https://en.wikipedia.org/wiki/Luminiferous_aether`.
[53] Wikipedia, Anthropic principle, (2021f), retrieved: 24 May 2021, `https://en.wikipedia.org/wiki/Anthropic_principle`.
[54] Wikipedia, New riddle of induction, (2021g), retrieved: 24 May 2021, `https://en.wikipedia.org/wiki/New_riddle_of_induction`.
[55] M. Polanyi, *Personal knowledge; towards a post-critical philosophy.* University of Chicago Press, Chicago (1958).
[56] T. Kuhn, *The Structure of Scientific Revolutions.* University of Chicago Press, Chicago (1970), 2nd edition, with postscript.

[57] Wikipedia, Witch-hunt, (2021), retrieved: 24 May 2021, https://en.wikipedia.org/wiki/Witch-hunt.

[58] D. Kahneman, *Thinking, fast and slow.* Farrar, Straus and Giroux, New York (2011).

[59] K. Burdzy and S. Pal, Can coherent predictions be contradictory? *Adv. in Appl. Probab.* **53**, 1, pp. 133–161 (2021).

[60] S. a. Cichomski and A. Osękowski, The maximal difference among expert's opinions, *Electron. J. Probab.* **26**, Paper No. 105, 17 (2021).

[61] R. Durrett, *Probability,* The Wadsworth & Brooks/Cole Statistics/Probability Series. Wadsworth & Brooks/Cole Advanced Books & Software, Pacific Grove, CA (1991), theory and examples.

[62] Wikipedia, Uncertainty principle, (2021a), retrieved: 25 May 2021, https://en.wikipedia.org/wiki/Uncertainty_principle.

[63] Wikipedia, Wave function collapse, (2021b), retrieved: 25 May 2021, https://en.wikipedia.org/wiki/Wave_function_collapse.

[64] Wikipedia, Double-slit experiment, (2021c), retrieved: 25 May 2021, https://en.wikipedia.org/wiki/Double-slit_experiment.

[65] Wikipedia, Schrödinger's cat, (2021d), retrieved: 25 May 2021, https://en.wikipedia.org/wiki/Schr%C3%B6dinger%27s_cat.

[66] I. Lakatos, *The methodology of scientific research programmes. Volume 1: Philosophical Papers.* Cambridge University Press, Cambridge; New York (1978).

[67] N. N. Taleb, *The Black Swan: The Impact of the Highly Improbable.* Random House, London (2008).

[68] Wikipedia, Image of God, (2021), retrieved: 9 May 2021, https://en.wikipedia.org/wiki/Image_of_God.

[69] D. Michaud, X. He, R. Peters, K. Jones and W. Wildman, Ludwig Feuerbach (1804–1872), *Boston Collaborative Encyclopedia of Western Theology* (2020), retrieved: 9 May 2021, http://people.bu.edu/wwildman/bce/feuerbach.htm.

[70] R. Nickerson, *Cognition and Chance. The Psychology of Probabilistic Reasoning.* Lawrence Erlbaum Associates, Mahwah, New Jersey (2004).

[71] Wikipedia, Christianization of Poland, (2021a), retrieved: 8 May 2021, https://en.wikipedia.org/wiki/Christianization_of_Poland.

[72] Wikipedia, Pagan reaction in Poland, (2021b), retrieved: 8 May 2021, https://en.wikipedia.org/wiki/Pagan_reaction_in_Poland.

[73] Wikipedia, Fountain (Duchamp), (2021c), retrieved: 10 May 2021, https://en.wikipedia.org/wiki/Fountain_(Duchamp).

[74] Wikipedia, Schrödinger's cat, (2021d), retrieved: 8 May 2021, https://en.wikipedia.org/wiki/Schr%C2%A8odinger%E2%80%99s_cat.

[75] Wikipedia, Twin paradox, (2021e), retrieved: 26 May 2021, https://en.wikipedia.org/wiki/Twin_paradox.

[76] Wikipedia, Epr paradox, (2021f), retrieved: 26 May 2021, https://en.wikipedia.org/wiki/EPR_paradox.

[77] Wikipedia, Book of job, (2021g), retrieved: 6 June 2021, https://en.wikipedia.org/wiki/Book_of_Job.

[78] J. Margolis, *The Truth About Relativism.* Blackwell Publishers, Cambridge, USA (1991).
[79] Wikipedia, Arbitrage pricing theory, (2021), retrieved: 26 May 2021, https://en.wikipedia.org/wiki/Arbitrage_pricing_theory.
[80] Wikipedia, Volkswagen emissions scandal, (2020a), retrieved: 10 May 2020, https://en.wikipedia.org/wiki/Volkswagen_emissions_scandal.
[81] Wikipedia, Tobacco Institute, (2020b), retrieved: 10 May 2020, https://en.wikipedia.org/wiki/Tobacco_Institute.
[82] Wikipedia, Noam chomsky, (2021), retrieved: 11 May 2021, https://en.wikipedia.org/wiki/Noam_Chomsky.
[83] N. Silver, *The Signal and the Noise: Why Most Predictions Fail - but Some Don't.* Penguin Books, New York (2015).
[84] J. Casarella, Paranoia, (2021), online; accessed 7 May 2021; https://www.webmd.com/mental-health/why-paranoid.
[85] L. Zmigrod, A psychology of ideology: Unpacking the psychological structure of ideological thinking, *Perspectives on Psychological Science* **17**, 4, pp. 1072–1092 (2022).
[86] Wikipedia, Creationism, (2021), retrieved: 27 May 2021, https://en.wikipedia.org/wiki/Creationism.
[87] A. Sokal, *Beyond the Hoax: Science, Philosophy and Culture.* Oxford University Press (2008).
[88] Wikipedia, Vaccine hesitancy, (2021a), retrieved: 20 May 2021, https://en.wikipedia.org/wiki/Vaccine_hesitancy.
[89] Wikipedia, List of probabilistic proofs of non-probabilistic theorems, (2021b), retrieved: 19 May 2021, https://en.wikipedia.org/wiki/List_of_probabilistic_proofs_of_non-probabilistic_theorems.
[90] Wikipedia, Transubstantiation, (2021c), retrieved: 10 May 2021, https://en.wikipedia.org/wiki/Transubstantiation.
[91] Wikipedia, Vincent van gogh, (2021d), retrieved: 8 June 2021, https://en.wikipedia.org/wiki/Vincent_van_Gogh.
[92] R. B. Semple Jr., Tate gallery buys pile of bricks - or is it art? (2021), New York Times, Online; accessed 15 May 2021; https://www.nytimes.com/1976/02/20/archives/tate-gallery-buys-pile-of-bricks-or-is-it-art.html.
[93] M. Heller, Theology for non-believers, *Mathematica Applicanda (Matematyka Stosowana)* **41**, pp. 249–254 (2013).
[94] R. Polt, *Heidegger: An Introduction.* Cornell University Press (1999).
[95] Wikipedia, Suspension of disbelief, (2021), retrieved: 9 May 2021, https://en.wikipedia.org/wiki/Suspension_of_disbelief.
[96] B. Skyrms, *Pragmatics and Empiricism.* Yale University Press, New Haven (1984).
[97] Good Judgment, Good judgment open, (2021), online; accessed 15 May 2021; https://www.gjopen.com/.
[98] UK Government, Unemployment, (2021), online; accessed 1 May 2021; https://www.ethnicity-facts-figures.service.gov.uk/work-pay-and-benefits/unemployment-and-economic-inactivity/unemployment/latest.

Index